DRILLING AHEAD

DRILLING AHEAD

*Tapping California's
Richest Oil Fields*

William Rintoul

Valley Publishers   □   Santa Cruz

ISBN: 0-934136-09-2
Library of Congress Card Catalog Number: 81-50167

Printed in the United States of America

Valley Publishers
A Division of Western Tanager Press
1111 Pacific Avenue
Santa Cruz, California 95060

For Susie and Jim

Contents

Acknowledgments

I would like to thank the reference staff of Beale Memorial Library, Bakersfield, and the staff of Taft Branch, Kern County Library, for help in making research materials available to me, particularly Nina Caspari, who recently retired after long service in charge of the Beale Library's Historical Collection, Mary Haas, who is in charge of the library's Geological, Mining and Petroleum Collection, and Elain Kanode, who is the Taft librarian.

Thanks are due many others who furnished information, pictures, and other help. I particularly would like to express my appreciation to George Austin, Betty Bean, Chet Davis, Jim Dorman, Bob Edgar, Bob Kilpatrick, Anne Loudon, Barbara Maxwell, Bob Montgomery, Mozelle Pollok, E.F. (Bud) Reid, Frank Rosenlieb, and Bob Wilfert.

William Rintoul
BAKERSFIELD, CALIFORNIA
OCTOBER, 1980

DRILLING AHEAD

1 Prophet of Cuyama

There was nothing to suggest in the modest appearance of the September 8, 1947, issue of the Taft, California, *Daily Midway Driller* that the six-page newspaper might harbor a bit of advice which, if heeded, stood a good chance to make the reader independently wealthy, and soon. If the advice had been printed in a respected business publication, or if the author had been better known, it might have carried some weight, but the *Driller* was not the *Wall Street Journal*, nor did James W. Travers have any special claim to fame as a financial advisor. The columns of the newspaper, which sold for five cents, were filled on that particular day, as on others, with the news that one might expect in a small-town paper, including such items as a prediction of forthcoming oil prosperity in the form of planned expenditures by nine companies of $6,260,000 to drill 203 wells, an announcement that the General Petroleum softballers would oppose the Tupman Merchants at Franklin Field in the first of a series of games to decide the community softball championship and an assurance that culture was not going to be forgotten on the West Side. A Chamber of Commerce committee headed by Jim Terry was making arrangements for the presentation of the play "Dear Ruth" by the Pasadena Playhouse at Taft Union High School Auditorium.

On an inside page beside advertisements for cotton dresses at Penney's for $2.79, smartest styles and sizes for all, and food specialties at La Frans Snack Shack, including the New Orleans Pore Boy Sandwich for 20 cents, there was an article by Travers, an 81-year-old semiretired newspaperman who had dabbled in oil operations for a number of years, giving the author's opinion of the past, present and future of the West Side oil fields. The tone was unapologetically optimistic. Travers recounted how when he started drilling a well some seven years before near Derby Acres at the north end of the Midway-Sunset field, which enveloped the city of Taft, "Knockers from areas

Norris Oil Company's Cuyama No. 2, shown on a rework, was the discovery well for Cuyama Valley's first oil field. (William Rintoul)

from one to three miles to the south said I would get no oil there because that territory had been drilled up nearly 30 years ago. I asked one fellow the depth they had penetrated and he replied around 650 feet. I replied that if necessary, I would go to China to get oil, for I felt sure some would be found below 650 feet. I got a good well at 1,250 feet with 250 feet of sand. Since that time other Johnny-Know-It-All knockers have been relegated to innocuous desuetude, as Grover Cleveland was wont to say when he had vanquished a foe and thought he had consigned him to depths below."

In looking at the future, Travers wrote, "There are thousands of acres of potential oil-producing territory on the West Side, extending from Blackwells Corner to the San Emidio Mountains on the south and to the far west end of the Cuyama Valley."

He advised readers to acquire land or oil leases in Cuyama Valley "and if financially able, drill for oil, but get the land at any rate and sit tight until another optimist comes along and offers a royalty and cash of the privilege for drilling on it."

The theme was a familiar one that friends had heard before from the small, frail-looking Travers, whose voice, rendered quavery by age, would rise in high-pitched intensity when he spoke of Cuyama's oil potential.

The area to which Travers directed readers' attention was a high, arid valley in the erratic, multihued mountains of the Sierra Madre range midway between the San Joaquin Valley and the coastal city of Santa Barbara. Nomadic Indians had named it Cuyama, an Indian word for clam, after the fossil clam shells they found there on their treks through the pass from the San Joaquin to the coast. Once a Spanish land grant, Cuyama had been settled by a handful of persevering homesteaders, who made a spare living raising cattle and alfalfa, sugar beets and seed potatoes.

The advice offered by Travers—to get a land position and wait for something to happen—did not call for the investment of a substantial sum, nor was it beyond the means of those of modest income. It was within the reach

Jimmie Travers predicted oil riches for Cuyama Valley. Above, Travers with his wife, Margaret, on the front porch of their home in Taft. (Taft Branch, Kern County Library).

of an average wage earner willing to forego a few pleasures on payday. Much of the valley was federal land. Under existing law, the land was classed as noncompetitive mineral land, which meant that any individual desiring to lease the land's fancied mineral wealth could do so merely by filing on the land, paying a $5 filing fee and a 50-cents-an-acre rental fee for a five-year lease. In return, the lessee would own all of the oil and gas beneath the land, if any, except for a one-eighth share retained by the government. In other words, for as little as $100 an individual could lease some 200 acres which could form the setting, according to the spacing of wells prevalent in San Joaquin Valley fields, for at least twenty wells, assuming there was something worth producing beneath the land.

There was one major stumbling block that might have given readers pause if they seriously considered following the advice offered by Travers. Wildcatters had drilled the Cuyama before. The valley had been a target almost from the earliest oil booms soon after the turn of the century in the neighboring San Joaquin Valley. The signs that brought the first oil prospectors to the San Joaquin Valley also existed in Cuyama Valley. There were oil seeps and interesting outcrops. As early as 1910, almost forty years before Travers offered his advice, oil men like Frank Hill, the Union Oil Company of California production chief who directed the battle against the famed Lakeview gusher, had known about Cuyama's seeps and sometimes stopped to look at them when passing through the valley. A live seep in the bed of Cuyama River had come to light in 1913 and had been duly reported in the same newspaper in which Travers' article appeared. The *Driller* said a local resident had stumbled on the seep and reported it was making an estimated half a barrel a day of oil. The find had sent a horde of men hurrying over Grocer Grade to Cuyama, where in less than a week's time they located more than sixty mineral claims in anticipation of the day when Cuyama would be the next big oil field. The Cuyama seeps had conjured up visions of an oil boom rivaling those that periodically broke in the San Joaquin Valley, but

unlike the latter valley, when wildcatters drilled in Cuyama, they came up empty-handed. Various companies and individuals had drilled without success and at least one major company, Standard Oil Comapny of California, had pronounced the valley a place whose geology would seem to preclude the existence of any major oil accumulations.

Advice was cheap, of course, and it's not known whether anyone followed the counsel that Travers offered in the *Daily Midway Driller*, except for one person. That person was James Wadsworth Travers himself. He purchased federal leases for 7,000 acres of wildcat ground, risking $3,500 of his money, plus filing fees, to back his judgment that Cuyama Valley held hope for major oil production. To make sure he got in on any fields that were discovered, he checkerboarded the leases through what he considered the most promising portion of the valley, which was that part near the western end in the vicinity of Chalk Mountain, a prominent landmark.

Wildcatters fanned out through Cuyama Valley, bringing in fast-flowing wells on spreads where homesteaders had eked out a precarious living. (William Rintoul)

Not all of Cuyama Valley's wildcats found oil. Richfield's Russell C No. 2 close by the Caliente Range went the deepest but failed to find production. The hole was abandoned at 12,981 feet. (William Rintoul)

Having taken leases, Jimmie Travers put the dream of Cuyama oil to the back of his mind and actively pursued another interest. It was the writing of a book he had envisioned for many years. Travers had an abiding interest in California history and those who had fashioned it. He was the vice president of The Society of California Pioneers and had, in fact, only a week before his article appeared in the *Driller*, attended the group's monthly directors' meeting in San Francisco in company with his wife, Margaret, and his close friend C.A. Pearson, a retired Taft groceryman, who with his wife had kindly driven the Travers to the San Francisco meeting. Travers was the son of California pioneers. His father had captained the sailing vessel *Hope* around Cape Horn, arriving in San Francisco to join the gold rush. His mother, May Estelle, had also been a pioneer.

May Estelle set out for California from her home in Portsmouth, New Hampshire, as a young lady of eighteen, accompanying a neighboring family whose daughter Genevieve was her closest friend. They sailed from Boston aboard the clipper ship *Columbia*. In Rio de Janeiro, Genevieve fell ill with fever. May Estelle and Genevieve's parents remained at the stricken girl's beside in the hospital. Genevieve died. May Estelle, too, was taken ill with fever. The Captain of the *Columbia* waited as long as he dared, but finally sailed, leaving May Estelle and Genevieve's parents behind. Two days later, the *Hope*, on its way to San Francisco, made port, battered from severe storms and badly in need of repairs. While the vessel was being re-outfitted, Captain Travers heard of the young American girl at the hospital and gallantly paid a call. It was love at first sight. May Estelle and Captain Travers were married. The Captain took his bride, and Genevieve's parents, aboard the *Hope* and contined the journey. They nearly lost their lives when the vessel encountered mountainous seas in the voyage around Cape Horn, but the Captain had the crew batten down the hatches, strapped himself to the wheel and brought the vessel through the dangerous passage. The Captain eventually beached the *Hope* in the mud at the foot of Washington Street in San Francisco, where the vessel served as the Travers' first home before the Captain built the 10-room house on the block bounded by Eighth and Ninth and Jefferson and Grove Streets in Oakland. In this house James Wadsworth Travers had been born 81 years before.

It had long been Travers' dream to set down the story of his father and mother's trip and to tell of the lives they had lived, adding observations from his own life, all of which would span the entire history of California from statehood to the state's approaching 100th anniversary. While Travers waited for the discovery that would prove up Cuyama Valley as an oil province, he spent long but satisfying hours writing the book which he titled *California*, with the subtitle, *Romance of Clipper Ships and Gold Rush Days*.

Even as James W. Travers relived exciting days of the past, others quietly were working along lines that would make his prediction of oil riches in Cuyama Valley come true. There were two forces at work. They represented the dominant elements of oil exploration, one being a major company approach, the other that of an independent wildcatter. The approaches were tailored to the character of the two parties that would make things happen. Though dissimilar in style and resources, they both featured strong-minded individuals willing to gamble on their faith.

They shared one conviction. Both believed there was oil to be found in Cuyama Valley.

The major oil company was Richfield Oil Corporation, a Los Angeles firm which had survived receivership in the dark days of the depression and come back under the leadership of a new president, Charles S. Jones, to make a major discovery in the North Coles Levee field in the southern San Joaquin Valley, 15 miles southwest of Bakersfield. The company was willing to flex its exploratory muscles in places others were not so willing to tackle.

Bringing in a big well at South Cuyama, scene of a major discovery by Richfield Oil Corporation. (William Rintoul)

Frank Morgan, a geologist with a strong belief that the way to find oil is to look aggressively for it, headed an exploration team that included men whose names would become famous among California's oil finders. The team included R. Stanley Beck, Donald Birch, K.M. Cook, Rod K. Cross, Drexler Dana, Tom W. Dibblee Jr., Paul H. Dudley, Cordell Durrell, Rollin P. Eckis, Tom Fitzgerald, Lesh Forrest, Mason L. Hill, Harold W. Hoots, Clifton W. Johnson, H. Allen Kelly, Joseph LeConte, Frank E. McPhillips, Manley L. Natland, James B. O'Flynn, John L. Porter, Sam Stasast, Frank B. Tolman, A.J. West and R.T. White.

Morgan, born in the mining town of Sonora in California's gold country and holder of a degree in geology from the University of California at Berkeley, had cut his teeth some twenty years before on the discovery of the Elwood field, which was on the coast, forty miles and a mountain range removed from Cuyama Valley. At Elwood, Rio Grande Oil Company, which later merged with Richfield, had leased wildcat ground on the basis of Morgan's recommendation and, with another company, Barnsdall Oil Company, had persevered with a prospect that others scorned. The result had been the discovery of a major oil field. The producing sand at Elwood occurred in the Vaqueros formation. Morgan had a hunch that Cuyama was a sleeper, and that the Vaqueros was the horizon that would put it in the pay column. He sent his geologists to Cuyama Valley and what they came back with convinced him that the valley, far from being the wildcatters' graveyard it was generally taken to be, in reality held the promise of significant oil production.

The on-the-ground survey conducted by Richfield's geologists was a careful, painstaking effort. One of those assigned to the study was Tom Dibblee, whose work in later recounting would become legendary through Richfield's exploratory ranks. Dibblee, traveling afoot, would drop out of sight for a week or more at a time, disappearing into Cuyama's wastelands to come back with precise drawings and notes from which geologic charts could be prepared. After one such trip into the field, he was said to have turned in an expense account of $14.92. Rollin Eckis, his superior, could not believe it. "Tom," Eckis said, you couldn't even get enough to eat for $14.92." Diblee reportedly replied, "Oh, I find lots of things to eat up in the hills."

As the reconnaissance went on, Richfield quietly began mapping mineral ownership and searching out the addresses of those from whom it would seek leases. Even as James W. Travers was advising readers of the *Driller* to place their bets on Cuyama Valley, Richfield in a budget meeting in the fall of 1947 was adopting a plan to drill its first exploratory well in Cuyama the following year. The decision was kept confidential. If there was any hint that a major company was planning to challenge Cuyama, others might jump in to take cheap protection acreage. Some of the desired land could be lost. At best, the price would go up. When Richfield moved in a rig, it wanted to reap the benefits, if any, not see them go to competitors unwilling to risk their own money.

In Cuyama Valley, events were moving to force Richfield's hand sooner than expected. While Richfield prepared to turn the resources of a forward-looking major oil company to the task of finding Cuyama's oil, three individuals with no less faith but considerably less finances were poor-boying an effort to find oil.

The spearhead of the group was an agile, silvery-haired oil prospector named George Hadley, who had spent nearly 50 of his 71 years looking for oil from Canada to Mexico. Some seven years before, Hadley, at the age of 64, had pulled into the valley in a beat-up car on which he subsequently would put more than 500,000 miles. Living out of the car, he had poked around looking for the clues that would put him on the trail of the valley's oil. It was a search that would have tried the stamina of a much younger man, but Hadley had persisted. He had become convinced that the valley was prime oil-hunting ground.

He had sought financial backing from an unlikely candidate for wildcatting honors. Halvern L. Norris, a scholarly middle-aged bachelor, lived in Ventura and interested himself in Oriental letters. He was a retired vice-consul to

Left to right, Charlie Jones, Frank Morgan and Rollin Eckis of Richfield Oil Corporation, three men who played major roles in the discovery and development of Cuyama Valley's oil. (William Rintoul)

Siam, which later became Thailand, and a veteran of many years' service in United States legations in Japan, China and Yugoslavia, among other countries. He had invested his savings in a small tool business and several houses in Ventura.

Hadley persuaded Norris to accompany him on a trip to Cuyama Valley, where he led Norris to a tiny water hole in the lee of the bank of the Cuyama River. There Hadley bored into soft earth at the bottom of the pool, stirring up a shadowy plume of live green oil. A handshake sealed Norris' participation. He agreed to put up $2,500, which proved only the beginning.

Hadley and Norris were joined in the Cuyama campaign by Arthur Scott, a Maricopa driller in his late thirties who had spent the greater part of his working life on the floors of drilling rigs. He was a lean, sunburned man, as quiet and mild in manner as his partners. Though he did not wager on his favorite card game, which was pinochle, he was willing to bet $7,500 of his savings and considerably more of his time on Cuyama's oil future, agreeing to take his share for handling the drilling work in stock in Norris Oil Company, which had no production to back its securities and little enough real property otherwise.

Along the way, Hadley by chance met Chester W. Colgrove, a veteran wildcatter who had a rig he wanted to sell, perhaps for an obvious reason. The rig had established the dubious record of drilling 17 dry holes in succession. Hadley struck a deal, paying cash and stock in Norris Oil for the rig. Later Colgrove also got a small sublease.

By the time Travers touted Cuyama's prospects in the *Driller*, the search for Cuyama production already had claimed two unsuccessful wells from the Hadley-Norris-Scott group. The first hole near the northern bank of the Cuyama River went to 1,755 feet, penetrating blue shales of the Morales formation before it was abandoned. The second in the same area cored oil-stained sands just below 1,700 feet, but when the well was tested, it flowed salt water, showing only traces of oil. The drilling campaign was a poor-boy operation all the way. There were times when the crew on the rig consisted only of Hadley and Scott instead of the five men who normally would have been there. The period was a difficult one for Norris Oil Company, which survived three reorganizations, including a time when some of the company's stock changed hands for only eight cents a share, with the proceeds going to pay hotel rent.

In November, 1947, some two months after Travers' advice appeared in the *Driller*, Norris Oil Company began another exploratory well close by Chalk Mountain, which Hadley long had believed to be the key to Cuyama's oil. The company had no staff of experts on whom to draw for advice and enlisted the help of Fred Sperber, a consulting petroleum engineer with an office in his home in Bakersfield. The year was drawing to a close when Norris Oil's Cuyama No. 2 on Sec. 25A, 11N-28W, San Luis Obispo County, cored a fine, silty oil sand from 1,874 to 1,925 feet in the top of the Santa Margarita formation. The excited partners bottomed the well at 1,973 feet and, following Sperber's advice, cemented a string of casing to try for production. On New Year's Day, 1948, Norris Oil Company completed the Cuyama No. 2 flowing 190 barrels a day of 21-gravity oil, cutting 0.2 percent water. It was Cuyama Valley's first discovery well. While the rate was not large, it did prove there was oil in the valley. One of those watching the operation was Jimmy O'Flynn, who was scouting the well for Richfield. Even before the well was completed, he realized that Norris Oil Company had a discovery. On New Year's Eve, which that year fell on a Wednesday night, he relayed the message to the home office that Norris had found oil.

In Los Angeles, Richfield's executives, like everyone else, had gone home for the day, anticipating a holiday on the following day that, depending on preference, might include the Rose Parade and the football game to be played that afternoon in the Rose Bowl pitting USC against the University of Michigan's Wolverines in a contest that Michigan would win by a score of 49 to 0. For many it was scheduled to be a long weekend, with work to resume the following Monday morning.

One of the innovations that Charles S. Jones had brought with him as president was something new in management technique. He had arranged for PBX operators to man telephones around the clock seven days a week to answer calls from the outside or from members of management, who in turn agreed to keep operators posted on their whereabouts. It was a system that had as its goal setting up a capability for quick reaction, and it was to pay off when the word came that oil had been discovered in Cuyama Valley.

When word of the discovery reached the switchboard, telephone operators went to work finding the members of the staff and summoning them back to the office. The company's landmen, under the direction of Frank McPhillips and A.J. West, along with lawyers and technicians, worked all that night, and through the following day— New Year's Day—and night as well, so that they would be ready to move on Friday morning, January 2. The company had a twin-engine Beechcraft airplane, piloted by Joe Brown and Charles Calvin. The plane was based at the Lockheed Terminal in Burbank. On Friday morning, the field was socked in. Brown managed to talk his way into getting clearance and took off through the soup. When the United States Land Office opened its doors that morning in Sacramento, the first to enter were landmen from Richfield, who filed on selected federal leases. From Sacramento, the plane carried landmen through five other stops that day, and a multitude in days immediately following, searching out landowners to get their signatures on leases. To speed up the process, the company bolstered the land force with salesmen experienced in meeting people and negotiating contracts, sending them out to get leases. When the long weekend ended on Monday, January 5, and Richfield's competitors opened their offices and started

Chalk Mountain, left, proved to be the site where Cuyama's first oil was discovered, as George Hadley believed it would be. (William Rintoul)

dusting off maps and old reports to get a land position in
Cuyama Valley, Richfield already had a broad land base. By
the middle of January, the company had acquired over
150,000 acres, representing in light of future events
almost 90 percent of the potential production in Cuyama
Valley.

Among those from whom the company sought oil rights
was James W. Travers. Richfield took on assignment from
Travers the rights for 1,000 acres, paying Travers a bonus
of $200,000 for the mineral rights he had acquired for 50
cents an acre, agreeing to pay an additional $400,000 out of
the oil they hoped to produce from the leased land.

All told, the whirlwind land campaign cost Richfield
more than $3 million of unbudgeted funds. The company
was staking its reputation and perhaps its future on the
belief that Cuyama would prove to be a significant oil
producing province. It was with understandable, if controlled, excitement that the company planned its first
exploratory well.

Before Richfield could spud in, the Cuyama discovery
well had begun to look considerably less attractive. The
well that Norris Oil Company had completed in the
shadow of Chalk Mountain faltered. The initial production
had not been high, but at least the oil had been clean.
Unfortunately, the well quickly began to go to water.
Within a month, the water cut rose from less than 1 percent to 50 percent. On balance, the discovery began to look
less like a major oil find than a small well that had opened a
field of uncertain potential. The well, in the opinion of
some, might even be taken as proof of the conclusion of
those who had stayed away from Cuyama on grounds the
valley's geology did not offer the potential for any major
accumulation of oil.

On April 23, Richfield spudded in with a Hoover Drilling
Company contract drilling rig from Bakersfield to drill the
Russell A No. 28-5 at a site approximately three miles
southeast of the Norris well. The Russell well ran high,
finding the Santa Margarita sand that Norris had tapped
for oil at a structurally high position. The sand was barren.

Richfield nervously drilled ahead. Eighteen days after spudding in, the company cored excellent oil sand at approximately 3,000 feet in the Vaqueros formation. Crews took the well on down to a depth of 4,218 feet.

On June 13, seven weeks after Richfield had begun its wildcat, a small group gathered by the well. Among those present were Charlie Jones, Richfield's president; Frank Morgan, the vice president in charge of exploration; and Hub Russell, Cuyama's leading cattleman, on whose Russell Ranch the well had been drilled. Russell later recalled that while he was not particularly excited, the 6-foot 6-inch tall Jones was very much on edge and paced up and down "like a chained coyote." It did not help matters when it seemed to require an undue amount of swabbing to bring in the well. In time, the well came in flowing 508 barrels a day of 38.2-gravity oil, cutting 1 percent water, from 390 feet of oil sand at 2,970-3,360 feet. The sand quickly was named the Dibblee sand, honoring the geologist who had played a major role in its discovery. The flow from the Russell Ranch well left no doubt Cuyama was an oil province.

Jones later recalled that Hub Russell seemed somewhat perplexed. "What does this mean?" Russell asked.

"Mr. Russell," Jones replied, "it means that you are a very rich man."

After thinking about this for a moment, Russell reportedly said, "Well, in that case I can buy me a good ranch."

Jones said later that he had thought many times since then what Russell's definition of a good ranch might be. His own definition, Jones said, would be one exactly like Russell's: 50,000 acres of fee land with the Cuyama River running through it and a major oil field in the middle of it.

Five days after the first well came in, Richfield scored again with a wildcat 2¼ miles to the northwest. The company's Anderson No. 37-30 came in flowing at a rate of 3,041 barrels a day of 33.5-gravity oil from an interval at 2,800-3,019 feet in the Dibblee sand. The New York Stock Exchange took note of what was happening. Overnight Richfield's stock jumped 10 points to $49 a share.

In the wake of success, Richfield's vice president in charge of exploitation, Dick Montgomery, put eight rigs to work drilling the area on ten-acre spacing. Within a month, the company was completing a well every four days. The burgeoning production posed a problem of how to get Cuyama's oil to market. There was a severe shortage of steel pipe. To solve the problem, Richfield bought thirty miles of battered and dented secondhand six-inch steel pipe that had been designed by Syd Smith of Shell Oil

Company for use by the Army during World War II. The Army had used the invasion pipe to string a surface pipeline across Europe to supply fuel for the tanks of General George S. Patton's Third Army. With the makeshift line, Richfield was able to start moving oil to market almost immediately by means of a tie-in with existing San Joaquin Valley pipeline connections. It was a temporary expedient that would be replaced in April, 1949, with a permanent 8⅝-inch conventional pipeline.

In Taft, Jimmie Travers was recovering from the effects of a severe cold that had put him in the hospital for the first time in his life at the very moment that Norris Oil Company was bringing in Cuyama's first well. By the time Richfield scored its first successes, he had gone back to work on his book. From the original concept of a book that would deal primarily with the stories his father and mother had told him, Travers had broadened the scope to include a potpourri of his own recollections and observations. He had added descriptions of fishing for trout as a youngster in Bear Creek near Santa Cruz, of visiting The Geysers in Northern California, where popular resorts catered to those desiring to take the baths, of attending the memorable prize fight between Jim Corbett and Australia's Peter Jackson and of his acquaintance with such well-known people as General Ulysses S. Grant, Mark Twain, Buffalo Bill Cody, Bret Harte, Jim Jeffries, Joaquin Miller and the actor Leo Carrillo. He had written of founding *Golden West*, a monthly magazine which for a time had been the official organ of the Native Sons of the Golden West organization; of experiences as a cub reporter in the 1880s on the *Oakland Enquirer*, a newspaper whose owner, Frank A. Leach, later became director of the United States Mints; and of his role helping Tex Rickard publicize the Battling Nelson-Joe Gans lightweight championship match when Travers was editor of the Tonopah, Nevada, *Daily Bonanza*. He had even compiled a collection of colorful names and what they meant for such California towns and places as El Tejon (the badger), Maricopa (Arizona tribe of Indians meaning "bean people"), Rio Vista (river view), San Luis

After discovery of oil on Russell Ranch, Hub Russell, center, liked to tell friends he got the crick in his neck from looking down to see if there was oil in his wells and looking up to see if it was going to rain for his cattle. Left to right, Joe Russell, Hub's brother and partner; Hub Russell; and Fred Sperber, consulting petroleum engineer on the Norris Oil Company discovery well. (William Rintoul)

Obispo (for St. Louis, the Bishop of Toulouse) and Yosemite (grizzly bear).

In August, work on the book took a backseat to something far more exciting. Richfield staked location for the first well on acreage it had acquired from Travers. The well was called the Margaret Travers No. 1 after Travers' wife. It was 1½ miles northwest of the discovery well that Norris Oil Company had completed. Richfield spudded in to drill the Travers well in September. The drilling bit found the Dibblee sand at 2,337 feet. Core analysis indicated the well should be a booming well. The company went in and tested. They recovered water. Geologists concluded oil had migrated out of the sand, leaving only a residual oil saturation. The company kept on going, drilling into the Soda Lake formation at 2,760 feet. There was nothing there. The hole was abandoned at 3,888 feet.

The failure of the well was a blow, of course, but Travers realized the Cuyama story was far from over. He wrote the dedication for his book. It read: "To the revered memory of my parents and to the other hardy and courageous pioneers and their descendants who are native Sons and Daughters of California, and who inherited the noble traits of fidelity, integrity and courage, this work—a true story, with a little fiction interspersed in a few places—is dedicated."

When he was not reading galley proofs, Travers found time to be an oil correspondent for the *Los Angeles Times*, sending news of Cuyama activity to Howard Kegley, whose daily oil column appeared in the *Times'* financial section.

There was no lack of news. Other companies quickly jumped into the play, among them Bandini, Barnsdall, Bishop, Federal, Gene Reid Exploration, General Petroleum, Gibson, Hancock, Honolulu, Humble, Independent Exploration, Pacific Western, Peak Oil Tool, Shell, Signal, Southern California Petroleum and Texaco. Hancock Oil Company became the first of the newcomers to complete a well, bringing in a good-looking producer the second time out on the Quality lease.

Oilworkers, attracted by wages of $16 a day or more for rig hands, flocked into the valley. Housing was at a premium. A tent city sprang up in cottonwoods near the booming Russell Ranch field. Before the discovery, Cuyama had never had a public telephone exchange. Soon there was a switchboard in Eugene Stutz's general store and pay phones in booths near the areas of greatest drilling activity. Postoffice business doubled from $7,000 to $14,000. The valley swarmed with speculators. When the rumor went out that land could be had for payment of delinquent taxes, eager speculators besieged county tax offices in Santa Barbara, San Luis Obispo and Kern County seats, demanding to see delinquent tax records for Cuyama districts. The situation got so bad that store proprietor Stutz suggested speculators wear carnations to keep them from trying to make deals with each other.

In the excitement no particular attention was paid to the news from San Francisco that the first shipment of Saudi Arabian crude had arrived at Richmond aboard the *California Standard*, a Standard Oil Company of California tanker. The vessel had left Ras Tanura on December 10, 1948, and arrived in January, after having taken the Pacific route, sailing east from the Persian Gulf. The 11,000-mile voyage avoided both the Suez and Panama Canal transits, which would have been necessary had the tanker taken the Atlantic route. Standard emphasized that the shipment was purely an experiment and did not in any way indicate plans in the immediate future to make regular hauls to the West Coast. No cost data was released, but speculation was that Saudi Arabian crude could be laid down in California for about $2.25 a barrel.

From Russell Ranch, Richfield moved five miles to the southeast to drill the Homan A No. 81-35 on Sec. 35, 10N-27W, Santa Barbara County. The wildcat was on a parcel that Glenn Homan, Los Angeles, had leased from the federal government for $170. Using an A.D. Rushing rig for the contract drilling job, Richfield took the Homan well to total depth of 4,392 feet, encountering nearly 300 feet of saturated Dibblee oil sand.

On May 4, 1949, Charlie Jones and other Richfield officials flew from Los Angeles to the firm's landing strip at Russell Ranch, accompanied by a group of Southern California newsmen there at the personal invitation of Jones. At the landing strip a chartered bus waited to take the party to the Homan drill site. The scene at the well was vastly different than the scene that had been played out less than a year before at Richfield's first Russell Ranch well. Jones escorted newsmen around the site, confidently answering questions and taking pictures of his guests with a camera that almost as quickly as the picture was taken produced a print. For many, it was the first Polaroid camera they had seen. The crew began to swab the well to bring it in. After several runs, the sand line started to unravel. The job had to be shut down, much to Jones' discomfort. The line was laid down and re-babbitted. The task took about two hours. Finally everything was ready. There was one more run with the swab. The well came in flowing at a rate of 5,000 barrels a day, proving up the largest oil field in Cuyama Valley.

Scouts mounted a round-the-clock watch on Superior's Government wildcat, which ventured into Morales Canyon to find more of Cuyama's oil. (William Rintoul)

22

With Russell Ranch producing some 20,000 barrels a day and new-found South Cuyama, now under intensive development, adding thousands more barrels, Richfield found itself with crude oil to sell. Shell Oil Company had the opposite problem. Shell was so short of light refining crude that it had to bring in several cargoes of Venezuelan crude to make up the deficiency. Richfield took care of Shell's problem. Before the year ended, Richfield also had agreed to sell Tide Water Associated Oil Company 5,000 barrels a day of Cuyama crude during 1950 and 1951, rising to 10,000 barrels a day during 1952 and to 14,000 barrels a day during 1953 and 1954. In a report to stockholders, Tide Water advised that the Cuyama crude had been analyzed in the Tide Water research laboratory at Avon and that tests indicated on account of its higher gravity and greater content of light components, such as gasoline and kerosene, Cuyama crude would yield a greater profit than the crude the firm had been obtaining from Fresno County properties of Mrs. Carrie Estelle Doheny and Los Nietos Company.

Though the Cuyama Valley search fanned out from one end of the valley to the other, oil development seemed always to skip the checkerboarded parcels that James W. Travers had leased from the federal government.

In late 1949, a printer at Wetzel Publishing Co., Inc., the Los Angeles publisher that was preparing to bring out the Travers' book, set the type for a brief paragraph that would be placed alone on the page that followed the book's dedication. The paragraph read:

"In Memoriam. It is with deep regret that we announce the passing of the author, James W. Travers, on July 22, 1949. The Publishers."

The book was published posthumously in 1950.

Shortly before, the Cuyama Valley oil boom came of age. In its December 31, 1949, issue, the *Saturday Evening Post* accorded national recognition to Cuyama with a feature article titled "Twenty Million Dollar Valley." John Burgan, city editor of the *Ventura Star-Press* and a successful freelance writer, introduced the *Post's* more than one million readers to Cuyama's excitement, telling of the men like George Hadley, Hal Norris and Arthur Scott, Charlie Jones and Frank Morgan who had helped to make Cuyama's boom a reality; and of ranchers like Hub Russell and Gene Johnson who had profited and their reactions to the turn of events, including Russell's dour contemplation, following the calculation of taxes, of the sign on the wall of his office which read: "No steer is so fat as one which scratches against the legs of an oil derrick." The author brought the story up to date on others like Chester W. Colgrove, who having come off 17 dry holes to strike it rich in Cuyama had organized a trust to repay in full the backers who had put up money for his unsuccessful ventures.

Just before 1949 ended, Superior Oil Company set events in motion for one last chance for the completion of a well on Travers acreage. The company moved northwest of the Russell Ranch field to drill an exploratory well in Morales Canyon, which was an area where Travers had taken leases. The company ran the well as a "no dope" hole, releasing no information on findings. When it was rumored there were oil shows, scouts from rival companies clustered on a nearby hill, mounting a round-the-clock watch. In April, 1950, Superior completed the Government No. 18-2 on Sec. 2, 11N-28W, San Luis Obispo County, flowing 399 barrels a day of 37.5-gravity oil from an interval at 5,620-6,125 feet in the Vaqueros formation. The following month, Hancock Oil Company brought in a well three-fourths of a mile southwest of the Superior well, completing Hancock-Oceanic No. 65-10 on Sec. 10, 11N-28W, San Luis Obispo County, for 68 barrels a day of 31-gravity oil from 1,710-1,920 feet in the Clayton zone. The well moved the Morales Canyon play to within three-fourths of a mile of land that Travers had leased and which his widow still held.

C.A. Pearson, Travers' long-time friend, decided it was time to try again. He teamed up with R.M. Steffen, a North Hollywood oil man, to back a wildcat on the Travers lease. Pearson's interest was motivated by more than the commercial desire to get in on Cuyama's oil riches. He hoped to bring in the first producing well on a Travers lease. He hired Fred Sperber to act as consulting engineer. In August, Pearson & Steffen moved in to drill No. 81-9 on Sec. 9, 11N-28W, San Luis Obispo County. The drill bit found the Dibblee sand at 780 feet. The sand was barren. Pearson & Steffen persevered, drilling through a thrust fault at 1,794 feet to plunge into the Morales, which was the formation that held the productive Clayton zone. The zone proved noncommercial.

Ironically, though James W. Travers had correctly predicted the Cuyama oil boom, the development itself had bypassed his leases. Even as Cuyama prospered, the last well to be drilled on a Travers lease was abandoned as a dry hole.

C.A. Pearson, a close friend of Jimmie Travers, made one last effort to bring in a well on a Travers lease. Above, Pearson with hieroglyphics on Painted Rock, a landmark on Carrizo Plains adjoining Cuyama Valley. The Dibblee sand occurs in the Painted Rock member of the Vaqueros formation. (William Rintoul)

2 Confusion Hill

The completion of a shallow well by Nelson-Phillips Oil Company in Placerita Canyon some 20 miles from downtown Los Angeles in April, 1948, occasioned passing interest in California oil ranks, mainly because the well, though hardly a barnburner, offered hope of being the first commercial well in a field that had been discovered 28 years before and only intermittently produced. There was no hint the modest well would set in motion a series of events that within a year would make Placerita the scene of a drilling boom some would describe as the most aggressive in California oil history, surpassing for a few frenzied months such earlier townlot drilling booms as Signal Hill, Santa Fe Springs and Huntington Beach.

Geologically, Nelson-Phillips' Kraft No. 1 on Sec. 31, 4N-15W, Los Angeles County, did not even rank as a new pool discovery. The well was drilled within the limits of the Placerita field, which had been discovered in 1920 by Equity Oil Company, a firm that no longer existed. The field was located at the western end of the San Gabriel Mountains about two miles east of the town of Newhall and seven miles east of Pico Canyon, where California's first commercial oil well had been completed in 1876. Pico Canyon had been the place where Pacific Coast Oil Company had gotten its start. The company was the predecessor of Standard Oil Company of California, the state's biggest producer.

Unlike Pico Canyon, Placerita had spawned no success stories. In fact, more than one operator had broken his pick. The field's oil was heavy crude, running as low as 11 degrees gravity. The first reported production did not come until October, 1925, more than five years after the discovery. The production was 6 barrels a day. Because of the low gravity of the oil, only 4 wells were drilled by the mid-1930s. Production from the wells ranged from 6 to 19 barrels a day.

One of the first rigs to go to work on "Confusion Hill" was this Clyde Drilling Company rig, which drilled for Rothschild Oil Company. (California Oil World)

Clyde Hall crew that worked the boom. Left to right: B.J. Spears, lead tong; Wilson Morris, cathead; E.A. Pitts, derrick; and Ray Gossett, driller. (California Oil World)

In 1935, another operator appeared on the scene with hopes of activating Placerita. Yant Petroleum Corporation, headed by M.R. Yant, had acquired the four existing wells and, in a further show of faith, purchased a parcel of as yet undrilled ground one mile to the north. The parcel was the north half of the northeast quarter of Sec. 31, 4N-15W. Before undertaking any drilling on the parcel, Yant had proceeded to drill a well on ground to the west of existing wells. He had completed Yant No. 5 from total depth of 2,735 feet, bringing in the well in September, 1936, for a disappointing 10 barrels a day of heavy oil.

Times were hard, and Yant raised money by subdividing and selling portions of the property he had acquired in Sec. 31, 4N-15W. He divided the parcel into small lots, some no larger than one-tenth of an acre. The descriptions in the deeds were by metes and bounds and were based on the assumption that the property contained 80 acres, as normally would be expected, there being no reason to believe that section lines were anything but true east-west, north-south boundaries. In the transaction, oil and gas rights passed to purchasers of the lots. However, Yant had the foresight to take a community oil and gas lease. Before drilling operations could be undertaken on any of the properties within the community lease, Yant Petroleum Corporation went bankrupt, one more victim of the great depression that had engulfed the country.

No further development was undertaken in the Placerita field until March, 1948, when Nelson-Phillips, a Los Angeles company, spudded in to drill the Kraft No. 1, using a standard steel derrick that had been erected several years before. The company took the well to total depth of 2,222 feet, cutting a fault in the process. The lower portion of the hole was barren, but the upper portion showed promise. Nelson-Phillips plugged back, cemented a string of 7-inch casing and completed the well from the interval at 585-718 feet in the Upper Kraft zone of Pliocene age flowing 70 barrels a day of clean 16-gravity oil. Though production was modest, the gravity of the oil was higher than previously encountered, and the depth was attractive. Successful completion of the well started a small drilling campaign. By the end of the year, 23 wells had been completed at depths ranging from 500 to 1,500 feet for initial productions varying from 25 to 175 barrels a day. Production from the pool averaged 450 barrels a day.

Though production at Placerita was small compared to that being developed in Cuyama Valley, the activity loomed large to at least one distant observer. M.R. Yant, who was living in Hollister, was following developments at Placerita with keen interest. Yant still owned interest in the north half of the northeast quarter of Sec. 31, 4N-15W, which lay only slightly more than one-half mile north of the Nelson-Phillips' Kraft well. He decided the time was ripe to try again.

Yant was working as an electrical contractor. The work brought him into contact with Ramon Somavia, a rancher in the Hollister area. Yant told Somavia of his prospect at Placerita and persuaded the rancher to put up money for an exploratory well. Somavia contracted with Gene Reid Drilling Company for a drilling rig, and in due course the Bakersfield contractor moved a small rig with a 96-foot Bender mast onto the proposed drill site on the brush-covered hill that Yant had subdivided in the 1930s. The subdivision, like so many others of its time, existed on paper only. The hill itself was empty, with no sign of streets or homes. Drilling began.

Just below 1,700 feet, the drilling bit found oil sand. The crew carried the well on down another hundred feet. The bit was still in sand. Somavia logged and ran casing. In early March, 1949, the Hollister rancher's Juanita No. 1 was completed flowing 340 barrels a day of 22.8-gravity oil through a ⅜-inch choke from the interval at 1,737-1,830 feet in the Lower Kraft zone. The well was the best yet for Placerita. It proved up an entirely new producing area with higher gravity oil than any that had been found before. Somavia lost no time hiring Jack Beckham, who had been with Clyde Hall Drilling Company in Bakersfield, to direct development of what was officially designated as the Juanita area of the Placerita field.

The discovery touched off a wild scramble for leases. The first finding on the part of those speculators looking for acreage was, not surprisingly, that the hill where Somavia had found oil—the Juanita area—had been highly subdivided, the area being covered by the north half of the northeast quarter of Sec. 31, 4N-15W. Some of those who had purchased lots from Yant had, in turn, subdivided their property, reducing parcels to as little as one-twentieth of an acre. The next discovery was that the assumption that the property contained 80 acres, in effect, that section lines were true, was not valid. In fact, the lines

There were 31 rigs working on "Confusion Hill" when Jerry Holscher took this aerial photograph. He made enlargements and sold them rig-to-rig. Bob Kilpatrick, working as a derrickman for Clyde Hall Drilling Company, was one of those who bought a photograph, paying $1 for the print. (Photo by Jerry Holscher, from Bob Kilpatrick)

were found to be off considerably, and the 80 acres that Yant had divided by metes and bounds were actually only 71 acres. Consequently, all of the parcels bordering on the exterior lines of the supposed 80-acre property were shortened by the overlap in deeds. Deeded parcels lapped over into adjoining lands and some on the north edge even extended up into Sec. 30, 4N-15W. Finally, it developed

that in many cases the legal titles were tangled. In some cases, as many as three claimants to the same tiny parcel appeared.

Obviously, the oil that lay beneath the subdivided area recognized no surface boundaries. The oil simply lay in the earth, waiting to flow to the surface in greatest volume to those who tapped it first, inevitably diminishing as the pool was exhausted. To those who got wells down first would go the flush production. Operators rushed to drill, mindful that each day's delay meant a loss of production to neighbors quicker on the draw. Scrapers graded roads, leveled drill sites, cut away settings for tanks. Contours of the hill disappeared. Portable spark-plug rigs crowded onto postage-stamp-size drill sites, working around the clock to put down wells as quickly as possible.

They called it Confusion Hill. The hill swarmed with drilling rigs, crews and service and supply men. One of the participants was Bob Montgomery, an engineer who had graduated from Stanford University two years before and, after a year with Richfield Oil Corporation, gone to work for Nelson Howard Company, a Los Angeles distributor of chemicals used in drilling fluids. Montgomery had handled sales at the Gene Reid rig that drilled the Somavia discovery well and, in the boom that followed, suddenly found himself living out of his car, handling sales and service to as many as 25 drilling rigs. He made an arrangement with Independent Exploration Company under which the company allowed him to pitch a surplus Army tent on property at the north end of the field for storage of chemicals, and hauled the 100-pound sacks from the tent to the various rigs on the fenders of his car. The job of servicing rigs was complicated by the fact that the narrow dirt road offering access to the oil field was constantly changing course. It was not unusual, Montgomery recalled, to start in to the field and find the way blocked by a grader, carving out a drill site in the middle of what had been the thoroughfare. It might be a matter of hours before the grader took the time to relocate the road for the convenience of those trying to get into or out of the field. In the course of his work, Montgomery became acquainted with Jim Casey, drilling superintendent for Miller & York, a contract drilling company operating several rigs in the booming development. The two became friends and four years later formed a contract drilling company of their own in Bakersfield.

At Placerita, contractors drilled the approximately 1,000-foot holes on footage basis, with the rate ranging

Portable rigs could drill and complete a Placerita well in as little as a week's time, sometimes bringing in the well for as much as 3,000 barrels a day. (Ira Carroll, Petroleum World)

from $6.25 to $6.75 per foot. The usual procedure was to drill the hole to completion depth before setting casing. In most wells, the operator landed either 8⅝-inch or 7-inch casing on bottom and cemented through perforations above the oil zone. A few wells had conventional water strings cemented above the zone. The operator demonstrated water shut-offs by perforating the water string above the cementing ports or casing shoe and then running a formation tester. This was normally done in one operation by running combination gun and tester. No particular difficulties or problems were encountered during drilling operations, and not a single well was lost because of mechanical problems. The average time required for drilling and completing a well ranged from seven to ten days, and the average cost from $27,000 to $30,000.

Wells tapped Pliocene pay some 300 feet thick at the most favorable structural position. The upper part of the zone was porous and showed excellent saturation while the bottom part was tighter and showed substantially less saturation. Initial productions ranged as high as 3,000 barrels a day. As long as semiflush production could be had, wells paid out in about three weeks. There was no gas cap, and operators produced wells at unrestricted rates, making no attempt to utilize the gas that was present to repressure the oil zone.

A concerned observer of the congested drilling boom was R.D. Bush, who as oil and gas supervisor headed the state's Division of Oil & Gas. The Division sought to enforce the provisions of the Spacing Act embodied in Sections 3600-3608 of the Public Resources Code of the State of California. The act provided that no more than one well be drilled per acre. With operators rushing to bring in wells on whatever size lot they could acquire, it was inevitable that the constitutionality of the act would be challenged. The challenge occurred in the case of People vs. Metcalfe Oil company. The state attempted to get an injunction to prevent the drilling of a well. On September 23, 1949, Superior Court Judge Clarence M. Hanson

Atlantic Oil Company completed Lockwin No. 1, marked by the Christmas tree, left center. A.G. McHale moved in to drill an offset, Woodworth No. 1, right, on a lease measuring 33 by 61 feet. Gordon Drilling Co. offset with Peggy Moore No. 10, rig on left. (Ira Carroll, Petroleum World)

denied the application for an injunction pending trial. The decision opened the way for an even more intense drilling campaign.

Confusion Hill boomed. Forty-eight rigs crowded into the townlot area. The emphasis was on making hole, Bob Kilpatrick recalled, and anything that looked like it could do the job was acceptable, even if it appeared to have been pieced together out of a Cherry Avenue, Long Beach, junkyard. Kilpatrick, who worked as a derrickman for Clyde Hall Drilling Company, said there was a chronic shortage of rig hands and on more than one occasion as the four-man crew was driving in to the job, they encountered one or another contractor's toolpusher standing beside the road, trying to hire the men before they got to their rig.

Because of the shortage, anyone who wanted to work another tour, as eight-hour shifts are called in the oilfields, could "double over," either on the rig where he normally worked or on some other contractor's rig. If a man felt like it, Kilpatrick said, he could work as long as he could stand up. Sometimes when the crew finished a tour, those who still felt like working would drive around the townlot drilling area, looking for a driller who was shut down because he did not have enough men to make a trip, that is, to pull drill pipe from the hole to change the bit or, if he was out of the hole, to run back in with the drill string. A floorhand could earn a day's pay for an hour or two's work helping the driller get back on bottom with the bit so the driller could make more hole. The pay ran around $12 to $13 a day for a floorhand, $15 to $16 a day for a derrickman and $18 to $20 a day for a driller.

Because so many rigs were jammed into such a small area, Kilpatrick said, pipe racks sometimes were so close together the crew had to look out or someone might reach over and use their pipe rack.

Once when the Clyde Hall rig was working on a hillside location a half mile from the nearest surfaced road, there was a heavy rain and the dirt access road to the drill site all but washed out. The crew was getting ready to cement casing, but they dared not try to bring the cement trucks

In drilling Meyers No. 1, Atlantic Oil Company found itself hemmed in on both sides, so the crew slid the kelly out along the runway. (Ira Carroll, Petroleum World)

up the hill, especially since the drill site was on fill and they already were having problems with the fill settling. They had the drivers park the cement wagons on the highway. The drilling crew fell to with thirty-sixes, that is, large pipe wrenches, and ran a pair of 2-inch lines from the well to the highway. It was nighttime, and the men stood with flashlights at intervals along the line, signaling when to pump cement and when to stop. They got the casing cemented.

Drilling crews, Kilpatrick recalled, like others, sometimes found the road to the field blocked. They might have passed over the road in the morning only to find when they started out at the end of the day's work that someone had moved in a drilling rig and was rigging up in preparation to put down a well on what only hours before had been the only road into the field. Once the crew ran into a more serious problem. They were almost to their rig when they found the way blocked by a man with a shotgun. The man claimed the company for which the contractor was drilling the well had no legal claim to the property. He refused to let the crew go into the changeroom to change clothes so they could go to work. The impasse contined for about an hour before a sheriff's deputy arrived. The deputy and the man with the shotgun talked awhile, then both left. The crew changed clothes and went to work.

One operator did not fare so well. Independent Exploration Company was drilling a well on what it thought was a valid lease when a deputy served an injunction, bringing the drilling operation to an abrupt halt. The oil company indignantly checked out ownership of the property. It developed that the party securing the injunction was, in fact, the legal owner. The company had cleared title to the lease through a title firm, but it turned out that a mistake had been made. The land actually belonged to someone else. Eventually the oil company collected $75,000 from the title company as compensation for the title company's mistake.

Peak production at Confusion Hill came early in October, 1949, when for a few days the field's wells produced 36,000 barrels a day, all but 450 barrels of it developed in only seven months.

Forty-one operators in a year's time completed 145 producing wells on the 71-acre hot spot that M.R. Yant had subdivided. The operators included several well-known independents and a host of newcomers to California's production ranks. Those who brought in wells were Arvin Oil Company, Atlantic Oil Company, B & J Oil Company, M. Barratt & Ernest W. Bysshe Jr., Bevo Drilling Company, Brayton-Phillips Corporation, Buffalo Oil Company, Camden Oil Company, Caravan Oil Company, O.F. Collinge, Holmes Oil Company Inc., Holmes & Everts & Associates, King Oil & Gas Company, Macliff Oil Company, Macmillan Petroleum Corporation, Mason & Wallace, Mawaco Inc., A.G. McHale, Warren L. Meeker, Tevis F. Morrow, Morton & Dolley, Newhall Refining Company, Northridge Oil Company, Len Owens Well Servicing Inc., Pacoil, Gus Pongratz, R.S. & L. Oil Corporation, Recknagel & Mangold Oil Company, Rothschild Oil Company, Serago Oil Company, Shaffer Tool Works, Shamrock Drilling Company, Louis V. Skinner & C.L. Best, Ray Smith

Oil Company, Ramon Somavia, Terminal Drilling Company, George Terry Drilling Company Inc., Three D Oil Company Inc., Twentieth Century Company, Watkins & Horton Inc. and Watmac Oil Company.

Others who brought in wells on adjoining leases included Crawford & Hiles, General Petroleum Corporation, Gordon Oil Company, Guiberson Oil Corporation, Independent Exploration Company, Indian Oil Company, Lake Oil Company, Midnight Oil Company, M & C Oil Company, Nelson-Phillips Oil Company and Standard Oil Company of California.

With operators producing wells wide open, production rapidly declined. During the same month of October, 1949, when production peaked at 36,000 barrels per day, decline set in and before the month ended, the field was down to less than 25,000 barrels daily. By the latter part of the month, most of the wells had ceased to flow. By the end of the year, most operators had put vacuum pumps on the casing at their wells to relieve some of the hydrostatic head and allow the oil to flow more readily into the wellbore.

General Petroleum Corporation, whose main trunk line from the San Joaquin Valley to Wilmington passed through the Placerita area, had the only pipeline serving the booming field, but the feeder line only moved a part of peak production. The greater part moved by truck and trailer units. Some filled up directly from the well without benefit of storage tanks. It was estimated that 100 truckloads a day left the field. Rothschild Oil Company was the largest operator in the Juanita pool and moved production to its two refineries near Santa Fe Springs as well as to the Olympic refinery near Long Beach, which Rothschild leased to process part of its production and store the balance.

Production from the Juanita pool demoralized the fuel oil market in Southern California. Marketers had to move the fuel oil after recovering the gasoline and other light hydrocarbons. Rothschild and others reduced fuel oil prices and consequently captured a large volume of fuel oil business from other marketers. They also captured business from Southern California Gas Company. Gas consumers found they could save money by converting to fuel oil. Several communities operating steam generating plants also reverted to the use of fuel oil.

Crude from the Juanita pool had a posted price of $1.32 per barrel, but many operators sold production for $1.07 per barrel, or 25¢ below postings. When crude was delivered to refineries, the price was $1.32, but this included delivery and transportation charges of 25¢ per barrel.

The Division of Oil & Gas took a dim view of the gas situation and notified operators in the field that unless steps were taken to eliminate the blow of gas, the Attorney General would be requested to take action. Meetings of operators were called at once. For a time it appeared that a voluntary curtailment plan would be successful. When it became apparent that no results were being obtained, R.D. Bush of the Division of Oil & Gas requested the Attorney General to institute action. The Attorney General proceeded to prepare a case. The Division charged 21 corporations and 19 partnerships with "wastefully causing and permitting natural gas and natural gasoline contained therein to blow, release and escape into the air."

In response to the Division's action, Judge Hanson issued an order requiring operators to show cause why a temporary restraining order should not be granted.

Negotiations between interested parties seeking an outlet for the field's gas met with failure. It was found that the gas had a high carbon dioxide content, ranging from 15 to 25 percent, which seemed to be increasing, and a low Btu. value, which made the gas unfit for domestic use. A study indicated the calculated gas reserve was too small to justify the erection of an extraction plant. It also appeared that a plant for the recovery of the gasoline alone was not feasible, because the gas had a low gasoline content, running to about 0.5 gallons per thousand cubic feet of gas. The gas wastage injunction action was placed off calendar in February, 1950, and no further steps were taken to stop waste.

Drilling in 1950 dropped sharply, falling off from the 253 wells that were drilled in 1949 to only 55 in 1950. Almost as quickly as they had come, contract drilling rigs moved out. By the close of the year, oil production had dropped to 12,233 barrels a day, far below the 36,000 barrels per day peak of the previous year. Quiet came to Confusion Hill.

When the rigs left, the hill that M.R. Yant had subdivided looked like a tank farm. (California Oil World)

3 The Paloma Deep Test

The group of men gathered in the conference room of The Ohio Oil Company's district office at 1801 Oak Street, Bakersfield, on Monday morning, September 28, 1953, was headed by George Sowards, Ohio's division manager, who had made a special trip up from Los Angeles. The group included virtually the entire management staff of the Bakersfield office, among them Bob Miller, the Bakersfield manager; B.H. "Curly" Doan, the drilling superintendent; Tony Tokash, the production superintendent; Rick Shoemaker, the district geologist; and Bart Emery, the district petroleum engineer. The purpose of the meeting, as indeed it was to be the purpose of much of what went on that day, was to see that everything went well that evening when Sowards would appear as the guest on the inaugural program of the L.W. Potter Trucking Company oil news on KERO-TV's Channel 10. The television station was just starting up. It was the second in Bakersfield, following close on the heels of Channel 29 in bringing television to the southern San Joaquin Valley. The program, scheduled for 15 minutes each Monday evening, would be the only regular oil news program on television in California and, as far as anyone knew, the only regular one anywhere in the United States. Among those present that morning in the conference room was Bill Potter, who owned the vacuum-tank trucking service that had agreed to sponsor the oil news program. The Monday evening program would follow Cousin Herb Henson's Trading Post, an hour-long country music program featuring Cousin Herb, Billy Mize, Bill Woods and whatever country music stars happened to be available between appearances at the Blackboard on Chester Avenue or at various country music bars on Edison Highway.

Sowards was the ranking Ohio executive on the Pacific Coast and a logical choice as guest for the inaugural oil news program. The program would feature a subject that was a natural—Ohio's KCL A No. 72-4, a well being drilled

The Ohio Oil Company's KCL A No. 72-4 became the deepest well in the world on August 20, 1953. The exploratory well was widely known as the Paloma deep test. (William Rintoul)

in the Paloma oil field 17 miles southwest of Bakersfield. As of a month ago, the exploratory well, widely known as the Paloma deep test, had surpassed the world's depth record and kept on going in a monumental search for deep oil or gas. In breaking the record, the well had caught the attention not only of the oil fraternity but also, through wide news coverage, of the general public. In fact, only a few days before the television oil news program was to be shown, a photographer from *Life*, the nation's most popular weekly picture magazine, had spent an entire day photographing the rig and crews. For one picture, he had appropriated a diamond core head that had been used in the well, driven off some distance, placed the core head on the ground and framed the drilling rig in the circular, diamond-studded opening.

Almost as soon as the well had broken the world's depth record, The Ohio Oil Company had commissioned a documentary film producer to make an in-house film, which was to be available for showing at service club meetings or on any other occasion when interest might be expressed. The 16-millimeter film included footage of the drilling rig, action on the rig floor, key personnel and, in general, a thorough look at the entire drilling operation. The film seemed an obvious selection for screening on the television oil news program, but because the program would only be 15 minutes long, it had been decided that only portions of the film would be shown, enough to run about five or six minutes. The oil company had sent the film to Bakersfield several days in advance so that those sequences to be shown might be selected and spliced. The splicing had been done by a newly formed enterprise seeking to establish a medium for handling television advertising and programming. George Sowards was to offer commentary as the film was shown, and therefore he naturally wanted to see the film in advance so that he would have some idea what to say as he watched it that night at the television station. Sowards' work for Ohio had taken him on many trips in and out of the country, and flying first class, he had on occasion found himself seated next to actors and television personalities, so he had no trepidation about appearing on television. His conversations had convinced him, though, that it was a good idea to be prepared.

To that end, he had decided to screen the spliced movie in the conference room and had invited others in the office in to see the film. Though someone in the Bakersfield office had had the foresight to secure a projector, it developed that no one had thought of a screen. The wall seemed inadequate, so Sowards, accustomed to making decisions on the management level, hesitated not at all, simply send-

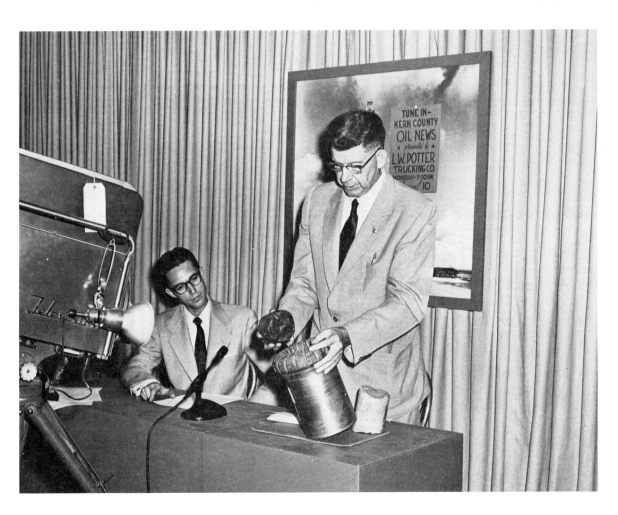

George Sowards, Ohio's division manager, demonstrated how the diamond core head worked while announcer Rudy von Tobel watched on the inaugural program of the L.W. Potter Trucking Company oil news on Bakersfield's Channel 10. (William Rintoul)

Driller Frank Sherman, right, accepted congratulations from Drilling Superintendent Curly Doan moments after the Paloma deep test drilled past 20,521 feet, breaking the world's depth record. (William Rintoul)

ing a staff member downtown to buy a screen. The oversight was the second slightly unsettling thing that happened that morning in connection with the planned appearance that night on television.

Before settling down to see the spliced film, Sowards in company with Bill Potter had made a call at the Channel 10 studio in the Truxtun Avenue wing of El Tejon Hotel. They had found the ground-floor studio a scene of mad confusion, with cluttered offices and hurrying people. When Potter had sought to introduce Sowards to one of the KERO-TV people, the functionary seemingly had misunderstood the nature of the program and the role Sowards was to play and drawn the assumption that the dignified, wise-looking Sowards was to be an answer man of some sort, fielding questions on camera that would be

This was the view of the world's deepest well from up the derrick. (N.S. Woodside)

telephoned in from the viewing audience, not only regarding the oil industry but, judging from the impression the functionary gave, on any other subject as well. It had been almost as if Sowards was to be the resident lovelorn advisor. This misunderstanding promptly had been cleared up, and Sowards and Potter had held a reasonably productive meeting with Ken Croes, the station director, and Dave Horowitz, who was the producer for the oil news program.

In time, the person sent for the screen arrived back at the office, the screen was set up, the film threaded and the conference room darkened. The opening sequence featured a close-up of the rotary table spinning on the floor of the company rig that was drilling the Paloma deep test. There was a collective, horrified gasp from the audience.

"They're backing off," Curly Doan, the drilling superintendent, said in amazement.

The rotary table, instead of turning to the right as it would do in normal drilling operations, was turning to the left, which meant that the crew would be backing off the string of pipe, leaving it in the hole. The spectre of a fishing job in the deepest well in the world was frightening. It would prove prophetic.

The film quickly was run back onto the reel, the projector stopped, lights turned on and a hurried look taken at the film by one of those present with experience in such matters. The examination confirmed what had been suspected. The technician in splicing the film had inadvertently reversed the opening footage, causing the rotary table to appear to turn counterclockwise instead of clockwise. The error was quickly corrected.

That night numerous employees of The Ohio Oil Company saw the Potter Oil News on a specially installed television set at the Bakersfield Inn, where they were attending a company dinner for the purpose of hearing about the new Thrift Plan.

The program went without a slip. Following the film segment, George Sowards answered questions by announcer Rudy von Tobel, describing the deep well as an example of the efforts to which the oil industry was going to provide the nation with needed oil reserves. Sowards had brought an expensive prop with him to show those watching the program, a diamond core head used to make hole at the deep well. He displayed sandstone and shale cores taken from the well, one from a depth of 20,937 feet. He vigorously banged the cores together to demonstrate their hardness. He even gave viewers a scoop by reporting that some three hours before the show the deep well had drilled past the 21,000-foot level, the first well in the world to break that barrier.

Driller Paddy Ryan was at the brake on the morning when the Tehachapi earthquake threatened to put an end to Ohio's deep probe. (William Rintoul)

The saga of the Paloma deep test had its beginning in the early 1930s, when Ohio entered into the original Paloma lease with Kern County Land Company, owner of the land on which the well was being drilled. At that time, Ohio, one of the oldest crude oil producing companies in the industry, tracing its origin back more than 60 years to the Trenton limestone fields of northwest Ohio and northeast Indiana, was primarily interested in shallower prospect zones. Drilling to a depth of 10,000 feet was still a curiosity. It had only been a few years before that a Miley Oil Company engineer at the American Petroleum Institute's annual meeting in Los Angeles had read a paper in which he stated:

"The rotary drawworks with the driller at the brake and throttle is gradually approaching its limit. The amount of weight to be given the pipe and the speed at which it should be rotated in a 10,000-foot hole will be too delicate matters for the ordinary driller to judge without further practical aid."

In spite of such pessimistic predictions, the original lease set the stage for the eventual drilling of the Paloma deep test. One provision of the lease, suggested by Geologist Bill Kleinpell and adopted by the land company, stipulated that in due course of time the lesee must drill a test of the Eocene sand or quitclaim rights to the prospective pay. The sand outcropped at Devil's Den, a blisteringly hot highway community 60 miles northwest of Paloma, but dipped to unknown depth under the valley floor. It was theorized that the sand might lie at a depth of 18,000 feet, or more,

Petroleum Engineers Don Everitts, left, and Clayton Stephens were in charge of the mud program that enabled Ohio to drill to record depth. Above, Everitts prepared to weigh a sample. (Robert J. Smith, from Marathon Oil Company)

Geologist Rick Shoemaker and Tom Wilson examined cores for clues to oil or gas. (Robert J. Smith, from Marathon Oil Company)

beneath the surface of the Paloma lease. Though the drilling of a hole deep enough to reach the sand was then unheard of, the terms of the lease foresightedly looked ahead to a day when such depths might be within reach of the drill bit.

As time ran out on the original commitment for the drilling of a deep test in the Paloma field, Ohio chose to drill rather than quitclaim. District Geologist Shoemaker described the target as the Point of Rocks sand of Eocene age, a prolific oil-producing horizon in other, shallower Kern

County fields. It was anticipated that a depth of about 18,500 feet might be sufficient to give the company an answer, though in the absence of any previous drilling at Paloma deeper than 14,400 feet, there was no way to be sure. The proposed depth was still short of the world's depth record of 20,521 feet set some four years before by Superior Oil Company's Pacific Creek Unit No. 1 in Sublette County, Wyoming. The Superior well had been abandoned as a dry hole soon after it reached total depth.

Ohio chose a drill site on Sec. 4, 32S-26E, Kern County, in the south-central portion of the 12-year-old Paloma field, where 84 wells had been completed from depths ranging from 10,650 to 11,440 feet for a production that then averaged 11,800 barrels a day. The company increased its already large land block in the Paloma district through lease of an additional 2,598 acres from Kern County Land Company.

Early in October, 1951, rigbuilders began erecting the 164-foot Lee C. Moore steel derrick that would be used for

the deep well. Because of the availability of supply, Ohio chose gas as a drilling fuel. In the early 1930s the company had drilled a wildcat test to 7,958 feet on the Paloma property and completed the well for 9.2 million cubic feet per day of gas from a depth of 5,200 feet. The gas zone proved limited, and only six producing wells were completed. All but two had been abandoned by the time the deep test was programmed. It was decided to use one to supply drilling fuel, maintaining the other as a standby source.

Drilling was routine during the first weeks after Ohio's KCL A No. 72-4 was spudded in on October 23, 1951.

Drilling Crew A, left to right, Jim Helton, derrickman; Leroy Finley, Bill McLaughlin, Alfred "Red" McKay, rotary helpers; Bill Gillett, driller. Wyatt Harris was the relief driller. The drilling foremen were R.I. Durham, Ed Fletcher and Jack Weaver. Ivan "Smoke" Seanor was the rig mechanic. (William Rintoul)

Drilling Crew B, left to right, Otis Roberson, S.R. Norton, V.R. Poehner, rotary helpers; C.A. Craig, derrickman; W.T. White, driller. (William Rintoul)

Company-owned drilling equipment, featuring a National 125 drawworks powered by four 350-horsepower Superior gas engines, costing $1,600 to $2,000 a day to operate, made rapid time, averaging more than 600 feet of hole a day in days that later would be remembered with a degree of nostalgia by the accountants whose job it was to tabulate expenses on a cost-per-foot basis. At 10,947 feet the drilling crews cemented the heaviest string of casing ever run in California—279 joints of steel pipe, 9⅝ inches in outside diameter, weighing 513,732 pounds. The purpose of the casing was to protect the fields's already-proven pay horizon.

Beyond the casing point, progress slowed abruptly as hard, compact shales and sandstones were encountered. Ohio put diamond cutting tools to work. The cutting face of each diamond bit and core head contained hundreds of tiny, black industrial diamonds from the Belgian Congo. Unsuitable for use in rings or other jewelry, the diamonds were set in what was believed to be the hardest alloy man had ever made—a mixture of tungsten, cobalt and nickel fused by exposure to 3,200 degrees of heat. The cost averaged $6,000 for each diamond bit and $5,500 for each diamond core head, with salvage value running to about 55 or 60 percent.

Five hundred carats of diamonds, average for each diamond cutting tool, on the end of a lengthening string of more than 11,000 feet of 4½ inch outside diameter, 17.35-pounds-per-foot steel drill pipe twisted relentlessly into the earth, pushing the Paloma deep test down at a rate of one foot an hour in shale and two to three feet an hour in sandstone.

At a depth of 13,818 feet, one of those happenings occurred which make the oil driller, perhaps more than any other man, realize that the concept of terra firma—the reassuring belief that one's feet are on something solid—is mostly myth. Far below the surface of the earth, the bit drilled into an ocean trapped in sand. Salt water gushed from the well.

Clayton Stephens, the division petroleum engineer, and Don Everitts, the petroleum engineer who was handling the well, knew that control of the salt water flow depended on the drilling fluid—the rotary mud—circulated through the hole during drilling operations to cool and lubricate the bit, wash cuttings to the surface and help support the walls of the hole. They increased the weight of the rotary mud from 11.3 pounds per gallon to 16.6 pounds per gallon by adding barite, a weighting agent. This killed the salt water flow, but the heavy mud began escaping into the formation. Tests indicated that the formation would take mud if it weighed more than 15.7 pounds per gallon; salt water would break in if the mud weighed less than 15.3 pounds per gallon. From that time on, engineers and derrickmen, the "mud men" on a drilling rig, had to keep a constant watch on mud weight, weighing untold samples of mud on a carefully calibrated scale.

Powerful engines ground on day and night, sending the drill string deeper into the earth. Diamond bits rotated on

Drilling crew C, left to right, H. Tudor, daylight extra man; H.L. Patkoski, derrickman; Carl Mead, Bert L. Calhoun, D.F. Young, rotary helpers; Frank Sherman, driller. (William Rintoul)

Drilling Crew D, left to right, N.S. Woodside, J.B. Thorp, D.O. Shrader, rotary helpers; N.C. Tuttle, derrickman; Fred "Paddy" Ryan, driller. (William Rintoul)

bottom as long as seven days without relief. The steel derrick, towering as high as a 15-story building, became a landmark in the Paloma farmlands.

Far below the diamond drilling bit, another powerful force was at work in the earth. At 4:52 A.M. on July 21, 1952, the earth began to tremble with a sharp, rolling motion. Driller Fred "Paddy" Ryan, a veteran of 30 years' rotary experience in California fields, was at the brake, drilling with bit on bottom at 14,310 feet. Thrown from side to side by the destructive magnitude of the earthquake, Ryan cooly kept hold of the brake, then chained it down while the rig was still shaking. A crewman started down the steps leading from the driller's doghouse and was thrown off by a tremor, falling eight feet to the ground.

The derrick swayed so badly that the drilling lines, supporting the more than 100-ton load of the block, hook and drill pipe, slapped against the sides of the rig. The top of the derrick made an arc judged to be from 15 to 20 feet. Fluid sloshed over the edge of the mud ditch.

If Ryan had let go of the brake during the sudden, sharp quake, the five-ton block and one-ton hook would have been free to crash to the derrick floor. Crew men below would have been killed or seriously injured. The deep well almost certainly would have been lost. Vital equipment would have been damaged, drill pipe jammed down the hole.

After the first heavy tremors ended, Ryan picked the bit up off bottom and rotated until the rig and equipment could be examined for possible damage. After a thorough inspection, drilling was resumed at 8:30 A.M., some three and one-half hours after the initial shock. Later, when damage from the Tehachapi earthquake was added up, it totaled 13 lives lost, more than $45 million property damage. The epicenter of the earthquake was pinpointed 14 miles southeast of the shaken but unscathed Paloma deep test.

Days turned into weeks and weeks became months at Ohio's exploratory well. Some of the men in the drilling crews planted a garden beside the rig, growing corn, tomatoes and other vegetables. On a wooden platform beside the sumphole, trays full of round-cut rock, resembling the stone pestles left by the Yokut Indians who inhabited the Paloma flatland years before the advent of the oil man, accumulated and finally overflowed onto simpler wooden racks set on timbers on the ground. These were the cores, columns of formation cut from the earth in lengths up to 56 feet by the diamond core heads.

Geologist Tom Wilson, who "sat" on the well, described core samples and sent them to core analysis laboratories to be analyzed for porosity and permeability, twin indexes of a formation's ability to contain and yield oil. Shale samples were sent to paleontologists, the specialists who worked to locate the Paloma deep test in terms of geologic age. In a core taken at 14,972 feet, paleontologists identified the microscopic remains of an organism that had lived 15 million years before the birth of Christ, when tiny shell fish inhabited the steamy inland sea that later became California's central valley. This find placed the hole in the Saucesian formation of Lower Miocene age, an indeterminate distance above the objective Point of Rocks sand.

Ohio persevered. At a depth of 18,734 feet, the company broke the California depth record. Drilling crews continued to make hole.

Temperatures at the bottom of the hole increased with depth. While the drilling crews perspired in summer temperatures of 100 degrees and more, the tools they controlled worked in temperatures increasing up to 334 degrees at 20,003 feet. Oil base mud was used as drilling fluid. The effect of the high temperatures was only slight on the oil base mud. It would have been disastrous on the water clay base mud commonly used in shallower wells.

In the early afternoon of August 20, 1953, Driller Frank Sherman stood by the controls on the floor of the big rotary rig, his eyes on the weight indicator, his right hand on the brake, carefully feeding off weight to the diamond bit that was cutting into the earth far below his feet.

At 1:20 P.M., Sherman looked at the depth indicator beside the controls and saw there the figures that told him the Paloma deep test was at 20,522 feet. As one of five drillers assigned to the KCL A No. 72-4, Sherman had just set a new world's record for deep drilling.

The Paloma deep test had gone deeper than any of the hundreds of thousands of wells drilled in the United States, deeper than any drilled anywhere in the world. The record, of course, was incidental. The purpose of the well was to find oil or gas. Sherman kept cutting at the earth, averaging a foot an hour in the hard shale that looked like sculptured black granite when the crews brought it to the surface in cores.

In a way, it seemed fitting that Sherman should have been the driller at the brake when the Paloma deep test shattered the previous record, set four years earlier by Superior in Wyoming. A veteran of 20 years' drilling experience with The Ohio Oil Company, Sherman had been the driller who spudded the Paloma well 667 days before—at 3 A.M. on October 23, 1951. He had worked on several of Ohio's first gas tests in the Paloma field in the 1930s and liked to show people the Indian arrowheads, mortar and pestle he had found.

When he walked away from the controls at the end of the daylight tour, Sherman lit a cigar and speaking like the veteran driller he was, loudly and close to his listener's ear, said about breaking the world's record:

"It was not a one-man job."

On the afternoon the record was broken, one photographer, anxious to get human interest shots, focused his camera on the depth indicator at 4 P.M. when the tour changed, expecting to get animated expressions on the faces of the new crew as the men came on the floor and saw the record had been broken. First of the afternoon tour crew on the floor was Driller Bill Gillett, who was returning to work after his vacation. He spoke to Frank Sherman and went into the driller's doghouse without looking at the depth indicator. The next two men walked past the indicator without paying any attention. Finally the photographer stopped a rotary helper and told him the well had set a record. The rotary helper looked mildly interested and continued on his way to the water cooler.

Though the drilling job had put equipment to the acid test, there had been surprisingly few breakdowns. One occurred the day the record was set. Late in the afternoon the crew that had been on tour when the record was broken started for Bakersfield in Derrickman H.L. Patkoski's car. Near Old River, 10 miles from the well, the car broke down. The crew called a taxi to take them to their homes.

After the Paloma deep test broke the world's record, visitors drove out to see the well. As luck would have it, Kern County's road crews picked that time to resurface the road near Conner Station that led to the well. Visitors had to bounce in on the shoulder of the road because the oiled section was completely torn up.

Among those who came to see the well were engineers from other companies, eager to learn what they could about the drilling of deep wells; members of several Desk and Derrick clubs, the vigorous new North American organization for women who worked in the oil industry; even foreign visitors, including an engineer and geologist from West Germany who wanted to see how Americans had drilled a well twice as deep as the deepest one in Europe.

Few visitors left the well site without a souvenir—a fragment of rock from the overflowing trays which held more than 5,000 feet of cores cut from the world's deepest well. Some cores found their way to the Bakersfield Chamber of Commerce, becoming souvenirs to be passed out on promotional trips to other cities. Many were sent out of state, some out of the country. At least one found its way into the hands of Juan Peron, then president of Argentina. A delegation of American oil men from Bakersfield's Independent Exploration Company, investigating the possibility of a concession in Argentina, presented the core as a testimonial to American know-how.

In the rush for cores, men working on the well held back, preferring to get their souvenirs from as deep as possible. "From the bottom ten," Engineer Don Everitts said. "I'll make a paperweight of it." As matters developed, this was to prove a heartbreaking impossibility.

On October 29, 1953, some two months after the Paloma deep test had broken the world's depth record, drillers cut what was to be the final core, a 25-foot core that took the hole to total depth of 21,482 feet. It took 20½ hours to cut the core with diamonds.

For a few hopeful hours in midafternoon of that lazy Indian summer day, it looked as if failure might give way to eleventh-hour success at the deepest well in the world. More than four miles below the floor of the big rotary rig, the diamond-studded core head seemed to move faster through the sediments that had been laid down some 20 million years before in the Miocene age. Could it be cutting a sand permeable enough to hold oil or gas?

Late in the afternoon, the driller finished cutting the core, taking the well almost 1,000 feet deeper than anyone had gone before. He took a strain to pull pipe from the earth. The earth held the drill string in a death grip. The driller pulled to 30 tons above the normal 155-ton weight of the drill string but failed to free the pipe.

After initial efforts to free the stuck tools failed, the crew attempted to back the pipe off at the safety joint above the stuck core barrel. Pipe backed off at 11,590 feet—the point where drill pipe itself was stuck—and efforts began to recover the fish.

Drilling crews ran drill pipe in and out of the hole with all the tools of the oil field fishing trade: an impression block consisting of soft lead that was nudged into the top of the fish to record an impression of how the lost pipe lay in the hole; a magnetic block to pick up loose metal; a junk basket with a catcher into which cuttings of steel could be flushed; and washover pipe, which cleaned around stuck pipe to a point where the stuck pipe could be shot off, enabling the crew to pull the lost pipe from the hole.

Each round trip in and out of the hole meant between 6½ and 11 hours of back-breaking work handling heavy steel tools. High up in the derrick, the derrickman racked each 120-foot stand of pipe. Before the fishing job was over,

Derrickman Jim Helton could chin himself one-handed with the arm he used to pull pipe into the rack.

At 12,027 feet, the washover pipe was stuck. Jarring for as long as 2½ hours with 40-ton blows failed to loosen it. The pipe had to be cut and recovered in agonizingly small pieces.

Drilling crews labored for 202 long days, a record in itself, to clean out the hole to 17,237 feet, which was considered deep enough for adequate testing of oil and gas shows encountered during the course of the drilling job. During the fishing job, crews made 290 round trips, running wire-line stuck-point indicators 33 times, string-shots 80 times, inside cutters 26 times, outside cutters 25 times, overshots 9 times, spears 33 times, washover shoes 84 times, mills 4 times, die collars 6 times and impression blocks 23 times. Rig time cost was approximately $237,000. Fishing tool rental was $125,000.

The legacy of the Paloma deep test was these cores, crumbling in the seasons as time passed by what, for almost three years, was the deepest well in the world. (William Rintoul)

On bottom, Ohio left 4,245 feet of drill string, at the end of which lay a diamond core head, valued at approximately $5,500 and the final 25-foot core no one would see.

After crews had cleaned out to 17,237 feet, the company proceeded with plans to run casing. As early as 1951, when initial planning for the deep well was in progress, the company had contacted casing manufacturers for recommendations on size and grade of casing suitable for running to 18,000 feet. Safety factors had not been satisfactory for any of the pipe available at the time. A manufacturer perfected a process for quenching and tempering high-strength joints of P-105 grade steel and fabricated some special 26-pound, 5½ inch casing of the same grade. Every joint was pressure tested at the mill to 80 percent of the minimum yield strength. In addition, the University of Illinois made 24 full section tension tests on specimens of the pipe.

At Paloma, crews ran casing to 17,222 feet in 17 hours using air tongs. The final indicator weight was 168 tons. To extend cement-setting time as much as possible, the company used ice to cool mixing water to 60 degrees.

The temperature of the hole—as high as 330 degrees Fahrenheit at depth—complicated the task of perforating the intervals to be tested. As a dress rehearsal, Ohio ran a 20-shot, simultaneous-firing bullet gun with powder but no bullets to 17,140 feet. Fifteen minutes after the gun reached depth, a loud, sharp report told the men on the surface the gun had fired spontaneously.

A test run with a jet perforator proved satisfactory. The crew ran a four-section jet gun to perforate the first interval at 16,740-17,135 feet. The crew could not get the gun to the desired depth. When the gun was pulled, examination proved it had fired, probably at 13,540 feet. No casing damage occurred.

It took almost 16 days to jet perforate the first interval. Gun failures occurred often, mainly because of collapsing screw-port plugs, leaking ring seals and prima-cord failures.

The company tried a bullet gun for the next interval from 16,465 to 16,735 feet. The gun not only perforated the desired interval, it also fired a stray shot that perforated casing at 12,064 feet. It took 38 days to locate and squeeze the hole.

In spite of difficulties, Ohio succeeded in perforating a third interval above 16,270 feet, a fourth above 15,488 feet and four more between 14,285 and 11,520 feet. Testing presented no particular problem once intervals had been perforated. Unfortunately, all zones tested nonproductive.

On December 31, 1954, The Ohio Oil Company abandoned the hole. The cost of the Paloma deep test came to $2,250,000, of which Ohio bore the brunt, paying about $1.9 million. Dry-hole contributions paid the rest. They came from Hancock Oil Company, General Petroleum Corporation, Lloyd Corporation Ltd., Western Gulf Oil Company and Texaco Inc., all of whom had acreage that might have been affected by a discovery.

Though abandoned as a dry hole, the Paloma deep test did not prove a total loss. Soon after the well was abandoned, engineers converted it to a water disposal well, salvaging the hole for service as a deep drain through which waste water from operations in the Paloma field might be returned to the earth.

In a paper presented before the spring meeting of the eastern district of the American Petroleum Insititute's Division of Production at the William Penn Hotel in Pittsburgh, Pennsylvania, on May 11, 1955, Bart Emery, Ohio's district petroleum engineer, said of the Paloma deep test:

"As a field laboratory, the well served a useful purpose. As a profitable exploratory venture, it was a failure."

4 A Changing Scene

On a stormy January night in 1957, a DC-3 airliner was carrying 18 passengers and a crew of three from Oakland to Los Angeles to make connections east when the plane's navigational instruments failed. Although Los Angeles was only 15 minutes away, the aircraft could not put down there. Driving rain and poor visibility had closed the city's airports for all but instrument landings.

The pilot turned north, looking for a break in the storm. The strong wind, he surmised, had probably blown the plane off course. What did the darkness below conceal? The ocean? Mountains?

He gingerly began a slow descent. At 6,000 feet he spotted a dim cluster of lights. With less than an hour's fuel left, he circled lower, hoping to find an airfield. But he was disappointed. As he tightened the spiral of his descent, the lights turned out to be those of a plant near a small town.

Then, off to one side, the pilot of the crippled plane saw something that gave him hope: the headlights of scores of automobiles. Could it be Highway 101, the main coastal route between San Francisco and Los Angeles? He flew lower and, to his amazement, saw the cars stop, turn at an angle and park in a row so their lights outlined what looked like a level strip.

Skillfully, the pilot made his landing approach, sideslipping through pelting rain to bring the plane and its passengers safely to earth.

"Where are we?" was the first question the shaken passengers and crew asked as they stepped out of the plane.

"New Cuyama, California," replied a driver of one of the cars. "When we heard you circling, we figured you must be in trouble."

None of the passengers or crew had heard of New Cuyama, but they could hardly be faulted. The town of some 750 people 100 miles northwest of Los Angeles was California's newest. It had come into existence in the wake

To make oil development at Sansinena more aesthetically acceptable, Union Oil Company of California put wellheads and other producing facilities in concrete-lined cellars, hiding them from surface view. (Union Oil Company of California)

of the Cuyama Valley oil discoveries. The town represented one of the proudest accomplishments of a changing oil field scene in California. The scene was beginning to include drilling rigs soundproofed to minimize disturbance to urban neighbors and drilling sites landscaped to look like parks.

The oil town of New Cuyama had a private airstrip that was used occasionally by small planes, but never at night since there were no lighting facilities. The lights that first attracted the pilot's attention were those of Richfield Oil Corporation's gas plant in the South Cuyama field.

When the townspeople heard the drone of the twin-engine plane circling in the storm, they hurried to do what they could to help. About 50 persons jumped into their automobiles and drove to the airstrip on the edge of town to use the lights of their cars as beacons in the storm. No one called them out. They simply felt individually responsible.

As a result, 21 persons who might have lost their lives lived to see their homes again, among them a young mother en route to Martins Ferry, Ohio, with her three-month-old baby, four homeward-bound servicemen and others who had listed as final destinations Endicott and West Haverstraw, New York; Dayton, Ohio; Chicago and Effingham, Illinois; South Portland, Maine; and West Haven, Connecticut.

The town whose residents responded to the plight of those in the airplane had come into being as an answer to the needs of those who had come to produce Cuyama Valley's oil. Many of those who brought in the valley's first wells had commuted to drilling rigs and production facilities 140 miles roundtrip from Bakersfield in the adjoining San Joaquin Valley. As more men came to work, Richfield, the leader in development of Cuyama Valley's oil resources, recognized the increasing need for a conveniently located community where workmen and their families could live.

Nobody wanted a company town, with a company store and all that would imply. Nor did they want a rough-and-tumble, haphazard collection of shacks and tents like so many oil boomtowns of the past. What people did want was a community in which individuals could own their houses and be proud to raise their children.

To help along the dream, Richfield purchased part of a dry farmer's grain field near the South Cuyama and Russell Ranch oil fields, leveled land and with the help of professional planners laid out a model community. The company provided water and sewer systems.

The first houses were priced in the $10,000 class, a category that the lowest-paid employee could afford. Terms called for nothing down except the first month's payment of $50 to $85, which went against the principal. For employees, Richfield guaranteed 30-year bank loans at three percent interest.

The decade of the 1950s had hardly begun before the first frame homes were set on cement slabs on the 70-by-140-foot lots that made up the townsite of New Cuyama.

The model community of New Cuyama owed its existence to oil, but the town was far different from boomtowns of the past. (William Rintoul)

Pleasantly landscaped houses on tree-shaded streets soon looked as if they had been there for years. New Cuyama had none of the rawness often associated with new towns. (William Rintoul)

The oilmen, accustomed to thinking in terms of mobility, hit on the idea of having houses constructed by a builder in Bakersfield specializing in a wide selection of wood-frame houses built on an assembly line for trucking to home sites. In short order, sturdy trucks carrying "Mobilhomes" were a frequent sight on the road from Bakersfield to New Cuyama, where houses were lowered to preset foundations along streets named for Cuyama pioneers like Hub Russell, the cattleman who had persevered through the valley's leaner days eventually to reap dividends from oil development, and the Cebrians, who had held the original Spanish land grant.

Though oilworkers were the main residents, others were welcome too, and the town's population quickly came to include ranchers and businessmen. The school district bought several of the homes and offered them at low rent as a fringe benefit to attract teachers. The town even attracted retired people, who came to take advantage of Cuyama Valley's 350 days of sunshine a year with clear, dry summers and mild, fog-free winters.

Soon New Cuyama had grown into an attractive community of 205 houses set behind green lawns and flower gardens on friendly streets shaded by elm, Modesto ash, silver maple and fruitless mulberry trees. The facilities available in the model community included a modern shopping center with meat market, grocery store; general merchandise store, postoffice, real estate office, beauty shop, barbershop, ranch and legal offices; a community civic activities center; a park with picnic facilities, athletic field

Use of a shaft-driven rotary table cut down noise inherent in the chain-driven type on the Gene Reid Drilling Inc. rig used to drill Universal's Curtis No. 1 in the Sansinena field. (William Rintoul)

and ball diamond; a motel, restaurant and service station; a county sheriff's office; a county fire station; a branch library; an elementary school and a high school with a swimming pool, tennis courts and a nine-hole pitch-and-putt golf course; a district ranger station; and a medical center with an ambulance. The ambulance was manned by two-man crews, driver and attendant, fully certified by the state and Red Cross. Eight volunteers, all unpaid, made up the pool from which two were always on standby.

Almost as an afterthought, the community even had a small jail by the sherriff's office. Through the years, it held one prisoner. One of the community's clubs at a party at the Cuyama Buckhorn, the local restaurant, raffled off a free night away from home. The winner, amid good-natured claims the raffle was rigged, learned that the free night away from home was a night in the town jail.

One of the big events in New Cuyama was the annual Halloween Carnival featuring chili, spaghetti and lots of merchandise booths, with proceeds going toward the $2,000 four-year scholarship awarded a graduating senior

On the lot of Twentieth Century Fox Studio, Universal Consolidated Oil Company discovered deeper pool production that would turn the declining Beverly Hills field into a major oil producer. Sun Drilling Company drilled the discovery well for Universal. (William Rintoul)

by the community's Cuyama Scholarship Foundation. Townspeople also sponsored an industrial arts scholarship fund, which made awards for advanced study in agriculture or technology. It was supported by proceeds from the annual presentation of the Cuyama Playhouse, a community theatre group.

Nina Lee Wade, who lived in New Cuyama with her husband Bill, a head wellpuller, and their three children, recalled a time when her sister-in-law came to visit from

Los Angeles. "I had to go to the grocery store, and she went with me. I said 'Hi' to the checker, then saw someone else I knew and said, 'Hi, Pete, how are the kids?' By the time we were out, I'd spoken to 30 or 40 people. My sister-in-law was stunned. She said when she went to the store, she never saw anyone she knew. She might go four or five times and never even have the same checker."

Among those oilworkers who brought their families to live in the town, four were unique, underscoring the success of the community. The four did not work in Cuyama Valley. They worked in San Joaquin Valley oil fields, preferring to live in New Cuyama even if it meant commuting 60 miles or more a day to do it.

Even as the planned community of New Cuyama was a far cry from the oil boomtowns of the past, what was happening in other California oil fields was just as amazing. Times were changing. Drilling rigs that would not have been recognized by an old-timer were at work developing oil fields that were landscaped to look like anything but oil fields.

One of the first efforts to change oil's image came in the Sansinena field near Whittier, where Union Oil Company of California faced a problem much more complex than simply finding oil. The problem was hardly one that Lyman Stewart, the company's cofounder and early president, could have foreseen when at the turn of the century he tramped over the hills east of Whittier at the edge of the Los Angeles Basin. To Stewart, the empty land obviously was oil country. He smelled crude oil. He saw it seeping from the hillside. Quick to seize the opportunity, he went to the owners of La Habra Rancho and purchased the mineral rights for some 3,400 acres designated as the Sansinena tract. Not until after he had made the purchase in 1903 did he bother to notify the company's directors that Union now was the possessor of a new block of wildcat ground. The company was strapped for funds, and the exploration department managed to drill only one hole, which was abandoned.

Inside Sun Drilling Company's soundproofed rig on the Twentieth Century Fox Studio lot, a crew set 7-inch casing to complete another well for Universal Consolidated Oil Company. (William Rintoul)

Union's wildcatters moved on to what they considered more promising prospects, and Sansinena languished until the early 1940s, when the exploration department decided to try again. In the meantime, subdividers in concert with the pressure of an exploding urban population had sold off the surface for small estates. The Sansinena prospect had been hidden by an aesthetically pleasing panorama of comfortable homes set in thriving avocado and citrus groves. The news that Union was planning to look for oil brought an angry reaction from those who had built homes and planted orchards. It brought forward visions of the oil developments that had boomed at places like Signal Hill and Santa Fe Springs, leaving a legacy of oil-stained derricks, tanks and pipelines. The homeowners initiated a campaign to push through a zoning law that would put Sansinena out of reach of the drilling bit.

Though Union still had legal title to the mineral rights, it looked like there might be a long drawn out legal battle to determine whose rights came first, those of the homeowners or the would-be wildcatters. Some of Union's geologists had second thoughts. They were not all that certain that there was oil at Sansinena. Some favored putting the prospect back on the shelf. One who wanted to drill was Cy Rubel, director of exploration. When the proposal was put up to Reese Taylor, who was then Union's president, Taylor responded: "Let's drill or get out." Those in favor of drilling prevailed.

For awhile, it looked as if the whole matter of mineral rights vs. surface rights might be academic. The first test was a dry hole. So were the second and third. It was not until the fourth try that Union hit. In May, 1945, the company completed Sansinena No. 15 on Sec. 30, 2S-10W, Los Angeles County, pumping 102 barrels a day of 22-gravity oil from the Puente formation of Miocene age in the interval from 3,346 to 3,591 feet. Completion of the well proved up production 1½ miles east of the nearest well in the Whittier field. Union moved one-quarter of a mile northwest of the completed well to drill Sansinena No. 17. Four months later the follow-up well came in for 156

A soundproofed derrick helped muffle the noise of drilling operations at Sansinena. (Union Oil Company of California)

barrels a day of oil from a pool some 700 feet deeper in the Puente formation. The two wells proved there was oil at Sansinena. They left unanswered the question of whether Union would be allowed to develop the field.

There were bridges to be built before the oil company could realize any return on the investment that Lyman Stewart had led it into more than 40 years before. The architects of one bridge were the company's landmen, who helped set the stage for development of Sansinena's oil by a unique "leasing" campaign. The company, of course, already owned the mineral rights. The landmen sought out homeowners and signed agreements with them committing Union to pay royalties on any oil that was produced.

There was the matter, too, of satisfying the Los Angeles County Planning Commission that Union's plans warranted a variance from the zoning regulations that had been adopted at the insistence of Sansinena homeowners. The company's strategy was simple. Union would develop the field from selected "islands" that would be hidden in canyons too steep or arid to be of use for other purposes. This involved borrowing a technique that had been developed in the 1930s to drill for offshore oil from onshore leases. The art of whipstocking a well, that is, directionally drilling the hole to put the bottom a predetermined distance away from the surface location, had been developed to a fine point at Huntington Beach. There companies had moved out into the ocean without, literally, getting their feet wet. Union proposed to do the same thing at Sansinena, directionally drilling wells from a single small drill site to tap the surrounding area of the field.

For homeowners who feared the noisy disturbance of drilling operations, the company agreed to sheath drilling rigs with soundproofing material.

After lengthy negotiations with the planning commission, Union in 1949 secured a variance to the zoning regulations permitting the company to drill 12 directional holes under its fee-owned property from a surface location to be confined within 150 feet of the Sansinena No. 15, the discovery well that had been completed four years before.

*Occidental Petroleum Corporation's
architecturally designed "oil derrick" at
Pico Boulevard and Doheny Drive won
an award from Los Angeles Beautiful,
a non-profit civic organization.
(Occidental Petroleum Corporation)*

Union lost no time beginning the drilling campaign, completing six directional holes before the year ended, getting initial productions ranging from 100 to 350 barrels a day. The company also obtained another variance to the zoning ordinance granting it permission to drill seven directional wells from the surface location of Sansinena No. 17, the follow-up well four years earlier.

The start that was made by Union proved to be the beginning for the drilling by the company and others of more than 200 wells, developing a production at Sansinena that peaked in 1956 at approximately 43,000 barrels a day.

One of the contractors that drilled at Sansinena was Gene Reid Drilling Inc. When the Bakersfield contractor moved in to drill the Curtis No. 1 for Universal Consolidated Oil Company, rig-up time took six days instead of the normal 16 hours for the same rotary outfit in other fields.

Rigbuilders spent four of the six days covering the derrick and rotary equipment with soundproofing material. The material consisted of two layers of vinyl-coated glass cloth with one-inch sheet fiberglass filling, especially heat-processed and quilted. Laps of the fire-resistant, washable soundproofing panels were securely fastened with three-inch safety pins. As a further safety measure, the laps were also wired.

Colored green on the outside to blend with the landscape, the soundproofing was bright yellow inside, bringing around-the-clock daylight operating convenience and safety for the drilling crew. On the outside, there were alternate orange and white strips at the top of the derrick as a warning to low-flying aircraft.

Inside exhausts were connected with two outside master mufflers through flexed tubing. The master mufflers were especially built to avoid any back pressure. The operation was set up so no fluid would be run out on the ground. There was no graded sumphole as there would have been in a San Joaquin Valley oil field. Instead, there was a square, 80-barrel tank to serve as a sumphole for shaker diggings. There were two mud tanks, one for clean mud, another for waste mud. There was a four-inch diaphragm-type pump to suck waste out of the cellar. Vacuum trucks were hired to haul away all waste.

Before preparations were completed to spud in the Curtis well, the approximate cost to Universal, the operator, was $30,000. The average cost of such preliminary preparations elsewhere in the California oil fields would have been about $5,000.

As development proceeded at Sansinena, the field became a showplace for demonstrating that oil field operations and relaxed living could go hand in hand. A casual visitor could drive through the area of homes and avocado and citrus groves without suspecting he was passing through an active oil field.

Once Cy Rubel drove a group of clubwomen through the field. The women had protested that oil fields ruined the landscape. Rubel offered to buy a new hat for any of the women who could point out an oil well. He finished the tour without having to buy a single hat.

In Los Angeles, the city council set the stage for an urban oil search by adopting a comprehensive zoning plan in 1950. Drilling, of course, was hardly new in the city. The first boom had come in the early 1890s with discovery and development of the Los Angeles City field. Various operators had drilled more than 1,100 shallow wells, including as many as four wells on a single 50-by-150-foot lot.

From the Los Angeles City field, wildcatters moved west around the turn of the century to discover the Salt Lake and Beverly Hills oil fields. Almost 400 wells had been drilled in the area that would later be occupied by such landmarks as Park La Brea and CBS Studios.

In those early boom days, the emphasis had been on getting oil out of the ground without regard to what neighbors might think of the forest of wooden derricks that threatened to engulf their homes. Aesthetics had not been an important consideration, nor was public sensibility. On one occasion an oil producer who found a buyer for a partially full storage tank had chosen to hasten the sale by simply opening the tank valve to empty out the crude, allowing the oil to flow down the street at the intersection of Glendale and Beverly Boulevards.

In contemplating a renewed oil search, the city's officials resolved to lay down ground rules. The zoning plan they adopted permitted the development of oil fields in residential and business districts, but only from approved drill sites. There was to be no more than one drill site for each 40 acres of mineral leases. The location of the drill site would be subject to the approval of the zoning administrator. The drill site was to be enclosed, landscaped and kept neat. The derrick and drilling equipment were to be fully soundproofed. Vibrations were to be controlled, and odors contained. Only electric power could be used. Delivery of materials would be limited to daylight hours. Waste material was to be hauled away to an approved disposal site. Strict safety requirements were to be enforced, and bonds posted. When drilling was finished, the derrick was to be removed, pumping equipment installed below ground level and oil transported only through underground pipelines.

With rules established, a search for oil began unlike any seen before in Los Angeles, or anywhere else in the world. For one thing, it took ingenuity to run seismic lines. At one place, under a wide grid of railway tracks, a seismic crew crawled into a man-sized culvert to place the recording equipment and record their "shots," between the train schedules overhead. At another site, as raw land was cleared for a new freeway, the seismic crews enthusiastically swarmed over it ahead of the construction crews and took their readings.

Continental Oil Company in gathering seismic data used a technique that required a setting late at night to avoid auto traffic and other city sounds that would throw off the readings. The company's geologists feared other oil companies would learn of their search and lease the land ahead of them. To throw off competitors, the company had its seismic crew pose as street cleaners in white coveralls.

Land plays involved as many as 50,000 separate leases. More than once those seeking leases were ushered off the premises at gun point. One landman was thoroughly drenched from the garden hose of an irate citizen who believed that oil companies were trying to defraud him.

But wildcatters got their chance, and they proceeded to prove up 200 million barrels of new reserves beneath an area lying in a great arc from downtown Los Angeles to the waters of Santa Monica Bay and, ultimately, in the San Fernando Valley. They discovered eleven new oil fields and found significant amounts of oil under three existing fields.

The new field discoveries included Boyle Heights, which was proved up by Richfield Oil Corporation in 1955; Cheviot Hills, Signal Oil & Gas Company, 1958; Las Cienegas, Union Oil Company of California, 1960; Los Angeles Downtown, Standard Oil Company of California, 1964; Sawtelle, Occidental Petroleum Corporation, 1965; Sherman, Standard Oil Company of California, 1965; Venice Beach, Mobil Oil Corporation, 1966; San Vicente, Standard Oil Company of California, 1968; Union Station, Standard Oil Company of California, 1968; South Salt Lake, Standard Oil Company of California, 1970; and Pacoima, Standard Oil Company of California, 1975.

There were new pool discoveries in the Beverly Hills field by Universal Consolidated Oil Company in 1954, Occidental Petroleum Corporation in 1966 and 1967 and Standard Oil Company of California in 1967. Buttram Petroleum Company found new pool production in the Salt Lake field in 1961. Signal Oil & Gas Company proved up new pool production in the Redondo Beach area of the Torrance field in 1956.

In some instances, there were ingenious approaches to the matter of camouflaging drilling and production "islands." A case in point was the Beverly Hills field.

The field was an old one that had been discovered so many years before—in 1900—that there was no record to single out the discovery well or tell what its production had been, though W.W. Orcutt was credited with the find. The field had attained peak production of 680 barrels a day in 1912. Drilling had ceased in 1925 after 32 wells had been completed. The field had declined until by the early 1950s there were only two wells left, putting out about 75 barrels a day. Through the first half century following its discovery, the field had produced only slightly more than four million barrels. The production had come from the Wolfskill zone of early Pliocene age at an average depth of 2,500 feet.

Standard Oil Company of California's Packard drilling structure blended so well with its surroundings that some visitors found it difficult to believe the structure hid an oil field. (Standard Oil Company of California)

Universal Consolidated Oil Company decided it was time to look deeper. The company leased the Twentieth Century Fox Studio lot and, after clearing the drilling project through the necessary agencies, contracted with Macco Corporation to erect a 136-foot steel derrick at a site near Olympic Boulevard. Rigbuilders covered the derrick with vinyl-coated glass cloth containing a filler of 1½-inch-thick spun glass. The drilling rig was a completely unitized electric rig. In November, 1953, Universal with Sun Drilling Company as drilling contractor spudded in to drill Twentieth Century Fox No. 1 on Sec. 25, 1S-15W, Los Angeles County. The company went to total depth of

Interior view at Packard drilling structure showed ample space to accommodate two big drilling rigs, trucks, mud logger's trailers, pipe storage and other necessary production gear. (Standard Oil Company of California)

8,562 feet, plugging back to 7,185 feet to complete in Miocene sand in the perforated interval from 7,088 to 7,185 feet. In February, 1954, the well flowed at a rate of 525 barrels a day of 24-gravity oil.

To develop the new production, Universal had Macco construct a drilling "island" 200 feet by 300 feet on the motion-picture lot, designing the drill site to accommodate 14 wells. Wells were spaced at 10-foot surface intervals and drilled from a continuous cellar 170 feet long. Macco surrounded the excavated drill site with a reinforced concrete retaining wall varying in height from 4 to 21 feet. The contractor built a combination field office and residence on the site, surfaced the working area with gravel and built a heavy-duty macadam surface approach road. When drilling was finished, the derrick was removed and the drill site screened with plantings and shrubbery, leaving no visible evidence of the operation inside the drill site area from any adjoining properties. Later, another similar drilling island was constructed on the Twentieth Century Fox lot approximately 1,000 feet west of the first island. Fifty-two wells were drilled from the two islands.

Occidental Petroleum Corporation joined the play and found oil sand 1¼ miles east of the first drilling island. To develop this new production, Occidental erected the world's first architecturally designed "oil derrick" on a drill site at Pico Boulevard and Doheny Drive. The company concealed the rig within what appeared to be a modern, ten-story office building. Occidental enclosed the drill site with a 12-foot-high flagstone wall and landscaped the whole with trees, shrubs and ground cover. Macco was the contractor, with Linesch & Reynolds the landscape architects, Thomas & Hopkins the derrick designers and Hobbs-Susnir the design firm for production facilites.

Los Angeles Mayor Sam Yorty officiated at ribbon-cutting ceremonies on January 20, 1966. Among those invited to attend were the approximately 3,000 residents of the area who as lessors would share in royalty payments. Mayor Yorty praised Occidental for the company's "outstanding contribution to civic beauty in a heavily-populated area of the city."

A month later at a luncheon in the Los Angeles Chamber of Commerce building, Robert S. Bell, chairman of the board of Packard Bell Electronics Corporation and an official of Los Angeles Beautiful, a nonprofit civic organization, presented an award to Occidental for "going well beyond the call of duty in spending approximately $1,000,000 to build the novel sky-blue derrick."

An official statement by Los Angeles Beautiful said: "Los Angeles Beautiful salutes this new approach by Occidental Petroleum Corporation to instill beauty of design into every change in our environment, including oil exploration and recovery."

Occidental subsequently drilled 58 wells from the site.

A later discovery by Standard Oil Company of California led to the construction by the oil company of a unique building on Pico Boulevard at Genesee Avenue in West Los Angeles, 1½ miles southeast of Occidental's "office bulding."

Designed to look like a modern office building, the structure, named Packard drilling structure, met design criteria so convincingly, according to one story, that two vice presidents of a rival oil company actually drove by the building three times before a city patrolman convinced them it was really Standard's "oil well."

The exterior of the 13-story building was finished in horizontal steel panels of beige color. The grounds were luxuriantly landscaped. Inside, the soundproofed structure housed two drilling rigs capable of handling 60-foot stands of drill pipe. The entire drilling operation was conducted indoors, including loading and unloading of trucks, cementing and logging. With the totally enclosed set up, it was possible to pull wells and do other remedial work at any time—around the clock—without having to bring in additional soundproofing and vapor control gear. An innovation was a viewing balcony from which residents of the neighborhood and visitors could see the work going on inside.

Standard drilled 80 wells from inside the structure.

Development of new production from the Universal, Occidental and Standard drill sites proved to be the biggest single factor in the urban picture. Thanks to the surge, the Beverly Hills field peaked at approximately 34,500 barrels a day in 1968, far above the previous peak of 680 barrels a day in 1912. The new pool discoveries proved up approximately 100 million barrels of reserves.

At Cheviot Hills, Signal Oil & Gas Company's geologists identified a potential oil field but faced the problem of finding a drill site in an area that included some of the Los Angeles area's most expensive homes, including those owned by John Wayne, Fred MacMurray, Nelson Eddy, Jeanette MacDonald and King Vidor.

Jim Wootan, the company's exploration chief, found what he considered an ideal drill site in a brushy ravine inside the Hillcrest Country Club's golf course. The exclusive country club was a fraternity of millionaires drawn from the worlds of finance, the professions and show business. The initiation fees alone ran into five figures. What incentive could Signal offer to cause the club to allow the company to drill for oil a hundred yards from the clubhouse?

Jack Benny, one of the country club's members, quipped "Perhaps if we sign up with Signal, we will be as rich as Bob Hope or Bing Crosby someday."

Actually, Signal found the directors of Hillcrest interested. The directors, too, were concerned with rising maintenance costs, insurance, taxes and other overhead. The oil company succeeded in getting permission to drill.

Signal also managed to get a lease on the nearby Rancho Park golf course, which was municipally owned.

To assure that drilling from islands on the two golf courses would be noiseless, odorless and inconspicuous, Signal's acoustical engineers worked with Hollywood's sound-stage experts to make the drilling derricks soundproof. Architect Henry C. Burge of the University of Southern California achieved the effect of making the derricks "inconspicuous, if not invisible" through the use of landscaping and color. He planted 60-foot-tall palms, which cost Signal $800 apiece, plus Canary Island pines to conceal the derricks from golfers, and heightened the camouflage by painting the jackets grass green at ground level grading to sky-blue at the top.

The final proof that a 9,000-foot well could be drilled without noise, odor or detectable vibration came during the playing of the 1958 and 1959 Los Angeles Open golf tournaments at Rancho Park. Highly strung golf pros played critical shots in the shadow of Signal's drilling rig without being disturbed by the fact, if they were even aware of it, that an oil well was being drilled just over the fence from the green.

Fifteen wells were drilled from the Rancho site, 33 from the island on Hillcrest.

At Las Cienegas, Union Oil Company of California and a partner, Signal, took leases from nearly 25,000 home and property owners. Union, as operator, undertook initial development after the discovery from an island in the parking area of a shopping center on Pico Boulevard near the intersection with San Vicente Boulevard. After drilling was completed, Union built a small, office-type building over the production equipment, enclosed the rest of the site with a brick wall and landscaped the property. The field became known as the only oil field with an address: 4848 West Pico Boulevard.

The site was one of five from which the Las Cienegas field subsequently was developed. One hundred and twenty-two wells were drilled from the five sites.

In the Salt Lake field, Jade Oil & Gas Company took over the development opened by Buttram from a drilling island near the corner of Beverly Boulevard and Fairfax Avenue in the vicinity of Farmer's Market. After drilling was ended, the company landscaped the site with tropical plants and flowering shrubs. Forty-two wells were drilled from the site.

At Venice Beach, Mobil Oil Corporation drilled and completed six wells to develop an offshore lease from a drilling island on the ocean front. The company camouflaged the drilling structure to look like a lighthouse.

At Redondo Beach, Signal Oil & Gas Company had to win a special municipal election to gain the right to drill from a waterfront site. Acting as operator for itself, Reserve Oil & Gas Company and Artnell Oil & Gas Company, the company used a soundproofed rig to drill and complete six wells, then landscaped the site to look like a park.

The site proved a mecca for sightseers. Visitors included the mayor and councilmen of Redondo Beach, officials of other Southern California communities faced with oil drilling offers and oil men themselves, including a petroleum engineer from Caracas, Venezuela, who wanted to see a landscaped oil field first-hand.

Encouraged by the response, Signal decided to put a large plaque at the site. Jim Wootan directed that the plaque be worded simply "so that a tourist from Iowa will know there are oil wells there."

Union shrouded drilling rigs in fire-resistant fabric padded with glass wool to soundproof drilling at Sansinena. (Union Oil Company of California)

5 Blue Waters of the Pacific

The pride that California's producers might have taken from their performance in 1953 was tempered by the sobering realization that, though they had pushed the state's production to a record height, they were losing the race. In a state whose largest city, Los Angeles, claimed the title of the nation's number one gasoline market, where people thought little of commuting long distances to work and the two-car family had long ago become reality, producers in spite of having pushed production past the one-million-barrels-per-day mark for the first time in the state's history—to a total of 367.3 million barrels for the year— were no longer able to keep up. Even as producers were setting a record, demand was outstripping supply. California, second only to Texas in the ranks of the nation's oil-producing states, was becoming dependent on outside sources for the crude oil that was the lifeblood of its energy system.

To make matters worse, it seemed to be getting harder and harder to find oil. There was a growing belief that the best prospects had been drilled, that more than sixty years of wildcatting had turned up the big fields and that what was left was the fate that had befallen California's gold miners, who had been reduced to sifting through the tailings left by hydraulic dredges, looking for the small nuggets that had slipped through.

It was not a bright outlook, but the evidence in the wake of the Cuyama Valley discoveries seemed to support the pessimistic conclusion. After South Cuyama, wildcatters had continued to keep busy, but they had found no more big fields in Cuyama Valley, nor had they done particularly well elsewhere. Humble Oil & Refining Company had found the Castaic Junction field 33 miles northwest of Los Angeles in 1950 and the Rosedale field seven miles west of Bakersfield in 1951. The Texas Company had discovered the Honor Rancho field 32 miles northwest of Los Angeles

The "steel island" at Rincon, foreground, was California's first offshore drilling platform. General Petroleum Corporation's Ferguson pier, background, helped point the way toward the seaward move. (Division of Oil & Gas)

in 1950. Richfield Oil Corporation had found deeper production in the Wheeler Ridge field 25 miles south of Bakersfield in 1952. All had been welcome discoveries, but when one added together the production from the three new fields and the deeper pool, it only came to about 10,000 barrels a day.

Gloom set in, underscored by the feeling that the time was running out for significant finds in traditional hunting grounds in the San Joaquin Valley, Los Angeles Basin and Coastal Basin. It appeared that if wildcatters were going to make a major contribution to the state's oil production, they were going to have to broaden their horizon.

An obvious prospect area formed a horizon that for many California residents was the most pleasing vista of the state—the Pacific shore. Unfortunately for wildcatters, it was largely off limits. The State Lands Act of 1938 provided that only tidelands being drained of oil—or threatened with drainage—by wells on adjacent lands could be leased. The law closed the door on any search for new offshore fields.

Though offshore prospects were off limits, wildcatters were casting interested glances toward the blue waters of the Pacific. Big discoveries in the Gulf of Mexico had amply proved the wisdom of Union Oil Company of California's Cy Rubel, who advised, "When you are hunting for oil, you have to shake loose from the idea that there is any particular significance to a shoreline, geologically speaking."

The rich strikes off Louisiana were propelling that state toward the number two position in the ranks of oil-producing states, surpassing California. But drilling the waters of the Gulf of Mexico was far different from the challenge that faced drillers in California waters.

In the Gulf, the Continental Shelf sloped gently into the sea so that it was possible to move out of sight of land without getting into water any deeper than 18 feet, as Kerr-McGee Oil Industries, Phillips Petroleum Company and Stanolind Oil & Gas Company had done at Ship Shoal, 10.5 miles offshore from Louisiana. There in 1947 the three-company group, drilling from a platform, had brought in the discovery well that opened the first big oil field in the Gulf of Mexico, giving birth to a new chapter in offshore oil development.

Off California, the Continental Shelf dropped off sharply going to depths of 200 feet or more within a mile or two of the shore.

It was obvious that drilling California's deeper waters would require the development of new technology. Some chose to begin preparing for the day when the state's restrictions might be lifted, opening the way for exploration of what came to be familiarly known as California's "whale pasturage."

Among those looking to the future were four companies that in the same year when production peaked and demand forged ahead of supply—1953—formed an engineering group to see what they could work out in the way of technology and hardware to drill for oil off California's shore.

The companies included Continental Oil Company, Union Oil Company of California, Shell Oil Company and Superior Oil Company. They chose to identify themselves by an acronym taken from the first initial of each company's name, becoming known as the CUSS Group.

Offshore drilling, of course, was not new to California. It had been off California that the whole business had started. A few miles east of Santa Barbara, between the Carpinteria and Montecito Valleys, a portion of coastal land had been sold in 1883 to a man named H.L. Williams. On the property, Williams, a member of a spiritualist cult in Santa Barbara, founded a colony named Summerland for cobelievers who wished to live far away from society's commercialism.

Four years later Williams dug two wells, both of which encountered oil sands. By 1895, Summerland had 28 wells. Ironically, the settlement intended for a retreat had been transformed into an oil town. The following year, in 1896, W.L. Watts of the California State Mining Bureau made a study of the region's geology. In his report, he wrote, "It is also evident that the oil yielding formations extend south into the ocean ... at low tide springs of oil and gas are uncovered on the seashore."

The following year, the first pier was built to serve as a base from which to drill the country's, perhaps the world's, first offshore well. The pier was the first of 14 that ultimately would be built at Summerland, including one 1,230 feet long.

The picturesque piers, bristling with stubby wooden derricks, made Summerland one of the scenic sights along the route of the Southern Pacific's coast line, which the railroad liked to describe as the "Road of a Thousand Wonders."

The field's oil sand occurred at an average depth of only 220 feet, which meant that it was not necessary to build a huge derrick. The small derricks on the piers required no more support than piles driven under each corner of the rig floor. Drillers started wells through a section of conductor pipe driven into the sandy ocean bottom to shut out sea water. Completed wells were pumped with jacklines. The wells produced an average of one to two barrels a day of 14-gravity oil that sold for up to eighty cents a barrel.

Piers proved difficult to maintain. The collapses caused by sea action broke off casing below the water level, admitting great quantities of ocean water to oil sands. Production was too small to justify the cost of extending piers into deep water, and the drilling of wells from piers ceased at about the turn of the century.

The next flurry of offshore activity came in the late 1920s on the heels of the discovery of the Rincon and Elwood fields. Pan American Petroleum Company discovered the Rincon field, eight miles northwest of Ventura, in November, 1927, bringing in the discovery well from an onshore site. Barnsdall Oil Company, in a joint effort with Rio Grande Oil Company, found the Elwood field, 14 miles west of Santa Barbara, in July, 1928, bringing in the discovery well on a bluff overlooking the ocean.

The two discoveries set the stage for the further development of offshore technology beyond the simple piers and shallow wells of the Summerland beginning. By far the greater amount of development took place at Elwood, where fast-flowing onshore wells stimulated an early move into adjoining tidelands.

The beach at Elwood was underlain with firm, hard shale and the overlying sand sometimes shifted, making wood piling less than satisfactory. Operators turned to steel piling to construct the piers from which wells could be drilled. To support heavy drilling equipment, they used reinforced concrete-filled caissons, including a center conductor caisson and corner leg supports. To conform with state requirements, the rig floor was constructed as a leak-proof apron with a central sump designed to keep fluids out of the ocean.

As piers stretched as far as 2,300 feet into the ocean, corner leg supports filled with concrete and reinforced with piling sometimes swayed under wave action. Operators began to use stabilizers—massive circular reinforced bodies of concrete—around the bottom of each corner column. This led to the development of the Collins Single-Leg Stabilized Foundation, which consisted of three

At Rincon, Richfield Oil Corporation moved one-half mile offshore into 45 feet of water to build California's second drilling island. (William Rintoul)

The construction of Monterey Island occasioned a law suit that pitted the City of Seal Beach against Monterey Oil Company. The oil company won the right to build the island as a base from which to develop Belmont Offshore field. (Exxon Company, USA)

members, including a single large central column, a large stabilizer around the column at the ocean floor level and a base for the derrick on top of the column.

The depression that began in 1929 and deepened in the early 1930s put a premium on finding ways to reduce costs, even if it meant going to sea without the security of a pier connected with the shore.

Indian Petroleum Corporation had a problem on its Shudde tidelands permit at Rincon. The first well, drilled from an onshore site, failed to find production, but the well did turn up geological information that made it look as if a well drilled on the seaward extension of the permit might be a producer.

The Los Angeles company in 1932 decided to forego building a costly pier in favor of simply building a portion of a pier, locating the structure at the site where geology appeared most favorable. The firm constructed a steel drilling platform in 38 feet of water some 1,700 feet beyond the end of the nearest pier. It was the first offshore drilling platform off California. The structure quickly became known as the "steel island."

Subsequently, three wells were drilled. The relatively small amount of oil they produced was piped ashore. In January, 1940, mountainous waves battered the platform. The structure went down. There was no loss of life, but equipment was destroyed and wells damaged. In abandonment, the steel island made one more bit of history. It was the first instance in which the placement of abandonment plugs was done by deep-sea divers.

Marine Exploration Company, which later became Monterey Oil Company, set the stage for the next step into the sea. The company had a state tidelands permit off Seal Beach. The city had an antidrilling ordinance, so the company had to drill from outside the city limits. The first two wells were unsuccessful. However, it looked like there might be something farther out.

In 1948, Marine Exploration, drilling from an onshore site in Long Beach one-half-mile inland from the water's edge, whipstocked a well almost two miles seaward—9,271 feet—from the derrick floor. The drilled depth was 12,180 feet; the vertical depth was approximately 5,700 feet. The well reached an angle of 83 degrees. Ernie Pyles, vice president of Monterey, said that if the company had not changed its technique by doing such things as slowing down the pumps, there were indications the well would have come back toward the surface past the 90-degree angle.

The hole encountered oil sand. The company attempted to complete on the pump but was only able to put the pump

7,000 feet down the hole, leaving it 5,000 feet off bottom at a vertical depth of about 3,200 feet. The well produced a consistent 30 barrels a day of clean 26-gravity oil.

Encouraged, Monterey, successor to Marine Exploration, decided to build an island from which to develop the indicated oil field. The City of Seal Beach claimed a boundary three miles out to sea. Monterey decided it had two choices: seek repeal of the city's antidrilling ordinance or start operations and let the courts decide whether the city or state had jurisdiction over the tidelands.

The company hired Healy-Tibbetts Corporation, San Francisco, as contractor and began building Monterey Island in 1952 at a site 1½ miles offshore from Seal Beach in 42 feet of water. The city brought a criminal action, charging Monterey—and Pyles with violating not only its drilling ordinance but also its building code.

The case traveled through the Superior Court of Orange County, the Southern District Court of Appeals in San Diego and the Supreme Court of the State of California, with a final ruling after 14 months that the city had no jurisdiction.

Monterey, with The Texas Company as its partner, resumed the task of building Monterey Island. Designed for 70 wells, the circular island, 75 feet in diameter, had an outer rim formed of interlocking sheet-steel piling driven into the ocean floor to depths of 15 to 20 feet. The interior was filled with rock and sand barged in from Catalina island.

To carry working loads, parallel reinforced concrete beams were installed across the island, each supported on a line of wood piling. A reinforced concrete floor rested on the beams and sand fill. A rectangular wharf 74 feet by 51 feet extended shoreward from the island. On the seaward side, large boulders were placed to protect the island from tidal action.

While Monterey was building the island, Brown Drilling Company, the contractor selected to handle the drilling work, marked out the exact dimensions of Monterey Island in the Lacey Trucking Company yard on Signal Hill. The contractor assembled within the outline a specially unitized National 50-A rig to be used for the drilling assignment, then moved the rig to an onshore site and drilled a well with the outfit set up exactly as it would be on the island. Drilling crews were required to stay in the "circle."

The first well was spudded on Monterey Island on May 22, 1954. It was a "no dope" hole. Communication with the mainland was through a submarine telephone cable connected to the public telephone system at Seal Beach.

Those working at the rig were instructed that when anyone called representing himself as being with Monterey, requesting information, the party was to be told to leave his name. The person at the rig would check to be sure the name was on a list in the doghouse, then call back at the number listed.

Management did not want to take a chance on someone calling out and saying, this is so-and-so, drilling superintendent or geologist in the Monterey office, How deep are you? What are you in? Those on the rig did not have to depend on recognizing anyone's voice, but the person requesting information did have to be at the number listed to get a report.

Some four months after spud-in, the first well came in flowing 300 barrels a day from an interval at 5,906-6,280 feet. The field was named Belmont Offshore.

After several wells had been completed, a second deck was constructed above the first to house the drilling mast, making it possible to continue drilling while production equipment was being installed on completed wells on the lower deck. Before the end of the decade, 41 wells would be drilled from the island and some five million barrels of oil piped ashore.

Though operators had managed to tap offshore oil from onshore sites and from piers extending almost half a mile into the sea, from a steel island and from Monterey Island, all offshore development still had one common link: it followed discoveries made from onshore sites.

Wildcatters could not seek offshore fields, but they were allowed to make seismic surveys and do other preliminary work, including the taking of cores, to identify prospective producing structures for the day when they might be able to explore them.

Inevitably, the need arose for the development of specially equipped boats capable of drilling in the open sea.

In 1950, Union Oil Company of California secured a 173-foot ex-Navy patrol boat, christened "Submarex," and mounted over-the-side a small rotary drilling rig, using a seismic drill-type rotary table mounted in a carriage with a fore and aft axis to compensate for the roll of the vessel. The rotary table was driven by an electric motor through an automobile-type transmission. The draw works consisted of a small construction-type hoist driven by a 25-horsepower gasoline engine.

During the next few years, several other ex-Navy boats were converted to over-the-side drilling, but they were only capable of drilling to shallow depths.

In 1955, Standard Oil Company of California commissioned Craig Shipbuilding Company, Long Beach, to modify for use as a drilling vessel an ex-Navy LSM (Landing Ship Medium), which already had been modified to serve as a lumber carrier. The 204-foot vessel's propellers were removed, technically making it a barge.

The conversion was unique. To permit drilling through the center of the vessel, a 10-foot diameter hole was cut completely through the hull, just forward of the bridge. A Wagner-Morehouse rig of a type normally used for 2,500- to 3,000-foot wells on dry land was centered over the "well." A torque converter was added to the diesel hoisting engines. The rotary table was gimbal-mounted, like a ship's compass, to offset rolling motion. Special joints of pipe with vertical slip sections, called "bumper subs," were provided to compensate for the heave of the vessel. Six such joints with 24-inch telescoping sections allowed 12 feet of up-and-down slack to keep the drilling bit on bottom as the ship rode the waves. The company christened the vessel *Western Explorer*. It had the capability to drill in about 300 feet of water.

About a month and a half after Standard started its hull through the Craig shipyard, Richfield Oil Corporation acquired a similar hull, an ex-Navy LSM which had seen post-war service in the Pacific Coast lumber trade as the *Jessie Andrews,* and modified it in the same way in the same yard, with only minor differences in rig construction.

As a measure of the company's avowed philosophy of sparing no expense in making the barge livable, an especially long bunk was installed for the toolpusher—W.P. "Slim" Burson of Bakersfield—who was able to fit his six-foot five-inch frame into a bunk 80 inches long—the maximum allowed by the length of his quarters.

The conversion of an ex-Navy LSM into a new class of drilling vessel was accomplished at Craig Shipbuilding Company's Long Beach shipyard by cutting a 10-foot diameter hole completely through the hull. Standard's "Western Explorer" was the first center-well drilling vessel, followed closely by Richfield's "Rincon," shown above during the conversion. (William Rintoul)

In the fall of 1956, a new center-well drilling vessel made its debut. The vessel was the *CUSS I,* the largest floating drilling vessel in the world. It was the outgrowth of three years of planning and work by the engineering group formed by Continental Oil Company, Union Oil Company of California, Shell Oil Company and Superior Oil Company.

The *CUSS I* represented an expenditure of approximately $3 million for the conversion of an ex-Navy YFNB freight barge into a vessel capable of drilling to depths of 10,000 feet or more in water depths up to 400 feet. The barge had a gross tonnage of 1,480 tons with a length of 260 feet compared to the *Western Explorer*-type vessel with a gross tonnage of 650 tons and a length of 204 feet.

The newcomer to the Pacific Coast's fleet of drilling vessels utilized a diamond-shaped caisson through the hull with a modified National rig mounted well above the upper deck. The diesel power units and mud pumps were positioned on the main deck.

Innovative features included underwater television for observing the wellhead and an ingenious mechanical horizontal pipe-racking device. The pipe-handling method eliminated the need for a derrickman. Pipe was pulled and racked in doubles, that is, two joints at a time, by a system of elevators and chain hoists. The pipe was racked in a slotted conveyor system which had space for 9,600 feet of pipe.

In the months that followed, the *CUSS I* proved that full-scale drilling was practical from a floating vessel. It handled assignments in water depths up to 1,500 feet, drilling holes as deep as 6,200 feet. It maintained position in all kinds of weather, cutting some 300,000 feet of hole during its first year of operation.

Even as deep-water technology was being developed, the state legislature was looking at the decline in production that had set in after the peak in 1953 and moving to open offshore state lands for exploratory drilling.

Offshore looked like the best answer to the state's sagging production. Joe B. Hudson, Monterey Oil Company

The first center-well drilling vessels mounted small rigs normally rated for depths of 2,500 to 3,000 feet at onshore sites. The vessels opened a new dimension in offshore drilling. (Atlantic Richfield Company)

geologist, told those at a meeting of the American Association of Petroleum Geologists in Los Angeles that a prize of more than 10 billion barrels probably lay waiting to be discovered in offshore portions of Southern California's Santa Maria, Ventura and Los Angeles basins. Hudson said offshore seismic and geologic exploration in the past decade had turned up "well over half a hundred structures" which had been "mapped in detail." He warned that a great part of the offshore oil might never be recovered in the foreseeable future unless state and industry leaders established laws, rules and regulations under which the offshore oil could be produced economically and competitively in world markets.

In Sacramento, Assemblyman Joseph C. Shell, a Republican from Los Angeles, introduced four bills to open the offshore door. The passage of the Cunningham-Shell Tidelands Act of 1955 gave wildcatters the chance for which they had been waiting. The act permitted the State Lands Commission to lease offshore lands without the necessity of proving drainage. It opened for possible leasing a stretch of California coast from Oceano, near Pismo Beach in San Luis Obispo County, south to Newport Beach, Orange County. The act allowed the drilling of wells from piers, filled lands or any satisfactory type of fixed or floating platforms. It set a fixed royalty rate of 12 1/2 percent for exploratory leases and gave the commission permission to set a fixed royalty of 16 2/3 percent or a sliding scale royalty beginning with that figure on proved oil lands. In all cases, competitive cash-bonus bidding was required.

The first lease was awarded in January 1957. It was a 5,500-acre exploratory lease offshore from the pioneer Summerland field. Standard Oil Company of California and Humble Oil & Refining Company won the lease at a competitive sale with a cash bonus of $7,250,600.

Some legislators opposed the Cunningham-Shell Tidelands Act, claiming the state would not get a proper share of revenue.

Soon after the Summerland lease was awarded, the State Lands Commission at the request of the state legislature suspended further leasing until the legislature could study the 1955 act and make any amendments it believed necessary.

In 1957, the act was amended to change bidding and royalty provisions. The amended act stipulated that the commission in both proved and unproved leases had to specify a sliding-scale royalty on oil production, commencing at not less than 16 2/3 percent up to a maxiumum specified in the bid. It allowed the commission to offer leases for the highest cash bonus or royalty bid. A series of sales followed, offering wildcat acreage on the basis of highest cash bonuses.

The opening of the state's offshore lands to exploratory drilling precipitated a marine drilling boom. Ten drilling and coring vessels of various sizes and types operated simultaneously off California, including the redoubtable *CUSS I.*

A newcomer to the fleet was a large mobile drilling barge, which could be towed to the desired drill site and set up on the ocean floor as a stationary drilling platform. It was the $2 million *Pacific Driller,* a 4,000-ton vessel built in New Orleans and towed through the Panama Canal to California.

Monterey Oil Company, acting as operator for a group that included Humble Oil & Refining Company and Seaboard Oil Company, put the rig to work off Huntington Beach late in 1956.

The barge on stilts measured 200 feet long by 100 feet wide and was 13 feet deep. It was equipped with a standard 136-foot derrick capable of drilling to 16,000 feet in water depths up to 90 feet. It had eight large caissons, each 6 feet in diameter by 195 feet long, which were lowered to the ocean floor. Two DeLong air jacks gripped each 6-foot caisson. The jacks were operated simultaneously by air pressure to lift the drilling platform any desired height above the water, usually from 20 to 40 feet depending on wave action. The barge also had a self-propelled, tread-mounted tractor crane for use in loading or unloading drill pipe and casing or to perform any other lifting operations. There was a special antipollution membrane to eliminate any possibility of contamination. A conveyor trough carried cuttings and waste mud to special hopper tanks, which were loaded onto barges by crane to be carried ashore for disposal.

On the various offshore vessels, men worked a seven-days-on, seven-days-off schedule, receiving free board and lodging during their working week. There were four meals a day aboard the ships, and those operating the vessels prided themselves on the quality and unlimited quantity of the food. For off-duty hours, there was generally television, plus fishing and swimming. The men were carefully selected, with marine experience desirable, as well as an ability to get along with others within the limited confines of the vessels.

The safety record aboard vessels was good, particularly with respect to actual drilling. William Rand of Submarex Corporation, Santa Barbara, who pioneered the technology of drilling and coring from floating vessels, said he believed the slight movement of the vessels constantly reminded the workmen they were engaged in a potentially hazardous undertaking, causing them to be safety conscious.

At Rincon, Richfield Oil Corporation moved one-half mile offshore into 45 feet of water to build California's

The "CUSS I" was the Queen. The converted Navy freight barge proved that full-scale drilling was practical from a floating vessel. (William Rintoul)

second offshore drilling island. Guy F. Atkinson Construction Company began building the $4 million island in February, 1957, and finished work in September of the following year.

The outer edge of the island consisted of huge boulders. Successive layers toward the center became smaller and smaller, and the inner core was filled with sand. On the seaward side, the island was protected by 1,100 concrete tetrapods, each weighing 31 tons. The tetrapods, shaped like the jacks of childhood play, were built at a Carpinteria work yard and barged to the island.

The one-acre island had a long concrete cellar with two rows of 34 conductors to provide for the drilling of 68 wells, if necessary. Facilities for dehydrating the oil and separating the gas were installed. Richfield used a regular 136-foot derrick for the drilling assignment, completing the first well in October 1958. The 2,648-foot well went on production making 77 barrels a day of 30.6-gravity oil from the Miley zone. The company kept one string of tools operating continuously, completing 46 wells by August, 1960, when drilling was halted pending further geological studies.

The 3,000-foot causeway that connected the island to the mainland was only wide enough for one-way traffic. In order to get above wave action, the causeway curved up toward the middle so that the two ends were not visible to each other. To avoid the possibility of two vehicles meeting and one having to back up as much as one-quarter of a mile, an electrically operated set of signals was installed. A traffic light which flashed green or red was set at each end of the causeway. When the driver of a vehicle approached either end and saw the green light, he knew the way was clear. However, before starting across the causeway, he would reach through the car window and turn a switch, making the light show red on both ends until he had crossed the causeway and switched the lights back to green.

At Summerland, Standard Oil Company of California, as operator for itself and Humble Oil & Refining Company, used the *Pacific Driller* to drill two coreholes on the offshore tract that had been purchased from the state in 1957. The coreholes found oil sand.

Standard built a stationary platform in the National Steel & Shipbuilding Company yard at San Diego, towed it at a speed of three knots for a distance of 210 miles to Summerland and installed the platform in 100 feet of water 2.2 miles offshore. Platform Hazel was the first platform installed in California that was constructed in a shipyard and towed to the site.

The basic platform consisted of a 110-foot square deck mounted on a tower 75 feet square and 170 feet high. The

The "Pacific Driller" arrived under tow at Long Beach harbor after a trip from New Orleans through the Panama Canal. Chained to the platform's deck were hugh caissons and air jacks. (William Rintoul)

tower was floated to the job site on the four big caissons which formed the bottom portion of the tower's legs, each 40 feet high and 27 feet in diameter. Each caisson was pressurized to prevent leakage and also ballasted with 90 tons of sand for stability. The deck was barged to Summerland for installation after the tower was lowered into place on the ocean floor.

Positioning the tower necessitated the services of a derrick barge with a 250-ton capacity boom, which was used to steady the tower and regulate the structure's descent as the buoyancy in the caissons was reduced. Once on bottom, the caissons were sunk 22 feet into the ocean floor by means of high-pressure water and air jets which literally hosed away the bottom sands, allowing the caissons to finally rest on hard ground. The final anchoring was accomplished by filling the caissons with 6,000 tons of sand and concrete.

Designed for 25 wells, the platform stood 50 feet above the water in order to be safe from storm wave action. The one-of-its-kind 162-foot Lee C. Moore derrick was designed so that two wells could be drilled at the same time, if desired. Oil and gas were pumped through submarine pipelines to adjacent onshore facilities.

A unique feature of the platform and sign of the times was the presence of a helicopter landing platform, which furnished a quick, modern method for transportation of personnel and supplies.

The total cost of building and installing the platform was about $4 million.

Standard selected Western Offshore Drilling & Exploration Company, Long Beach, as drilling contractor. The contractor, drilling with a National 80-B rig using electric power transmitted from shore by a 16,500-volt submarine cable, took the Standard-Humble Summerland State No. 1 to total depth of 7,531 feet. The well in November, 1958, flowed 865 barrels a day of 36-gravity oil, proving up Summerland Offshore field.

The discovery represented a significant breakthrough. The previous push into California's tidelands had been in pursuit of discoveries made from dry land. The find at Summerland Offshore was the first new field discovery made from the sea itself, opening a new era for offshore development off California.

In lease sales that followed amendment of the Cunningham Tidelands Act in 1957, the state sold 31 tracts in the Santa Barbara Channel for bonuses of $169.5 million.

The discoveries that followed included Alegria Offshore oil field, Richfield Oil Corporation, 1962; Caliente Offshore gas field, Standard Oil Company of California, 1962;

Platform Hazel in Summerland Offshore field opened a new era for offshore development off California. (William Rintoul)

Carpinteria Offshore oil field, Standard Oil Company of California, 1966; Coal Oil Point Offshore oil field, Richfield Oil Corporation, 1961; Conception Offshore oil field, Phillips Petroleum Company, 1961; Cuarta Offshore oil field, Texaco, Inc., successor to The Texas Company, 1959; Elwood-South Offshore oil field, Richfield Oil Corporation, 1965; Gaviota Offshore gas field, Standard Oil Company of California, 1960; Molino Offshore gas field, Shell Oil Company, 1962; and Naples Offshore gas field, Phillips Petroleum Company, 1960.

To develop the fields, operators installed six new platforms. They drilled some wells from floating vessels and completed the wells on the ocean floor. In some instances, they drilled and completed wells from onshore sites.

At Conception Offshore, Phillips Petroleum Company converted the mobile barge *Pacific Driller* to service as a permanent drilling platform, installing it in 105 feet of water. The company renamed the platform "Harry" in accordance with Coast Guard regulations that required all permanent platforms in a particular region to be designated by a name starting with the same letter.

To see what effect platforms and drilling islands might have on fish life, California's Department of Fish & Game conducted monthly fish counts in the waters surrounding drilling structures, using scuba-equipped marine biologist divers. In 1961, the Department issued a report that stated, "After careful observation for two years, no deleterious effects have been discovered, and it has become obvious that oil islands and platforms are providing new homes for thousands of fish."

Richfield Oil Corporation made the first ocean-floor completion off California in March, 1961, at Rincon, completing State 1466 No. 102 in 55 feet of water three-quarters of a mile offshore for 75 barrels a day of 29-gravity oil from the Miley sand. The 2,290-foot hole was drilled from the *Venmac,* a barge owned by Western Offshore Drilling & Exploration Company.

Richfield encased the wellhead in a steel housing, and divers made the final connection. To prevent leakage,

there were double seals throughout and a valve designed to close automatically unless held open by hydraulic pressure in the control line. There was also a storm choke in the tubing string designed to close automatically with a pressure drop. The ocean-floor completion was the third in the world. The year before, Peruvian Pacific Petroleum Company, a subsidiary of Richfield and Cities Service Company, had completed a well on the ocean floor off Peru. The second completion had been made off Louisiana by Shell Oil Company.

While oil companies were opening a new chapter in California's offshore development, the *CUSS I* was selected for the initial testing phase of Project Mohole, a government-sponsored project to drill through the earth's mantle.

A new company had been formed early in 1959 to operate the vessel and offer the engineering and experience of the disbanded CUSS Group. Personnel of Global Marine Exploration Company included engineering and operating personnel lately associated with the CUSS Group, including Bob Bauer, president; A.J. Field, vice president and general manager; and Hal Stratton, vice president.

Drilling first in 3,500 feet of water, the *CUSS I* began to shatter the old belief that drilling could not be carried out in the deep waters off the Continental Shelf.

On March 30, 1961, the vessel, while drilling in basalt in 25-mile-per-hour winds and waves in excess of 12 feet, retrieved cores from 200 feet beneath the sea bottom in 11,700 feet of water near Guadalupe Island off the West Coast of Baja California.

Another highlight of the vessel's career occurred in late 1965, when she established a deep water drilling record by drilling, running and cementing multiple strings of casing without diver assistance in 632 feet of water.

Through the years, the vessel worked for most of the major oil companies. For years, she continued turning to the right up and down the coast of California and Oregon and in Alaska's Cook Inlet.

Fully assembled and on location, the "Pacific Driller" became a platform as sixteen 300-ton-capacity air jacks gradually lifted the 4,000 unit out of the water. (William Rintoul)

While *CUSS I* was in Cook Inlet, the State of Alaska offered a bonus of reduction of royalty to the operator whose rig could penetrate commercial oil sands the fastest. The race pitted *CUSS I* against the *GLOMAR II* and a WODECO vessel. The *CUSS I*, pulling doubles instead of triples, won the race.

After a long and noted career, *CUSS I* was retired from further drilling service in March, 1978, and dismantled for sale at Long Beach.

For those who had worked on the ship, the vessel was something special. An anonymous *CUSS I* crew member left the following note tacked to the galley bulletin board, containing sentiments shared by many.

"On March 22, 1978, on Pier D in Long Beach the *CUSS I* was quietly retired. To the men who worked on her she was known as *The Queen*. We were proud to work on her; she may have been small but she always did her job. She may have been a bit slower and her equipment may be outdated and not as efficient as her younger sisters. But you have to understand she is getting old and tired so maybe she should be retired to live her life out as a memory to the many men who served on her.

"On this last day of her duty she sits against the dock, the men who were left to clean her out were quietly going about their duties, not much was being said, for most are thinking can it really be? She was such a lady, sometimes moody as ladies will be, but she always did her job. There's a lot of sweat and blood on her decks and patches on her hull, but to us she is still *The Queen*. She will always be out there somewhere drilling.

"As I walk toward my car on the pier, not wanting to look back, for I want to remember her as I knew her: proud and beautiful. I say to myself: 'Farewell Queen, long live The Queen, *CUSS I*.'"

California's Department of Fish &
Game conducted monthly fish counts in
waters surrounding drilling structures.
SCUBA-equipped marine biologist
divers found oil islands and platforms
attracted thousands of fish to what
formerly had been marine deserts.
(Texaco Inc.)

6 The Case of the Disappearing Land

The war had ended in 1945, as everyone knew, but the Navy had an increasingly uncomfortable feeling that some unseen foe had drawn an invisible target over Southern California's Long Beach Harbor and placed the Navy's shipyard almost precisely in the center.

There were strange things happening at the shipyard, which had been built during World War II at a cost of some $75 million but, by the time the 1950s had rolled around, represented a replacement cost estimated at $170 million. The walls and foundations of buildings had begun to crack. Sewer lines had broken. Storm drains had ruptured. The railroad tracks that served the yard had begun to buckle.

It had not taken long to trace the mysterious disturbances to a strange phenomenon that was taking place at Long Beach Harbor. When surveyors made the annual August survey of Long Beach Harbor Department bench marks, they found that elevations were less each year than they had been the year before. One bench mark near Navy's shipyard in one year alone—1951—dropped 2 1/5 feet. The inescapable conclusion was that the land in the harbor area was slowly sinking beneath the sea.

When Navy had built the Long Beach facility, the shipyard had been six feet above high tide. The base was one that the Navy regarded as virtually irreplaceable and strategically vital to the nation's defense. It was the only one on the Pacific Coast which fronted on the Pacific Ocean. Bases at San Diego, San Francisco, Portland and Seattle were in bays or well inland from the ocean proper.

As harbor land sank, Navy like others with facilities in the area took what remedial actions were necessary to protect installations, doing such things as trucking in landfill, raising foundations and constructing dikes. The subsidence had continued until by the 1950s Navy's shipyard lay below sea level, protected from flooding only by dikes, some as high as twenty feet.

Stainless steel pump was lowered into a water source well for steady delivery of salt water to be used in flooding to halt subsidence and increase oil recovery at Wilmington. (The Port of Long Beach)

The situation had become so alarming that there was beginning to be talk about abandoning the yard and moving the Navy's operations to some undisclosed site somewhere else on the Pacific Coast. This kind of talk was enough to send chills through the business community of Long Beach. The annual civilian payroll at the shipyard was $37 million with 6,400 employees. The yard accounted for approximately one-seventh of the Long Beach area's economy. Also, general Navy activity would be cut if the yard were closed. There were 140 Navy ships and 40,000 uniformed personnel based in Long Beach, bringing an estimated $127 million into the area's economy. If the shipyard were moved out of the area, it would be a severe blow.

Navy had its own considerations, though. It was dependent on Congress for funds to operate the shipyard, and Congress was beginning to grumble about appropriating the huge amounts of money necessary to keep the Long Beach facility in operation. The cost of dikes soon had run past $7 million, and there seemed no end in sight to what might eventually have to be spent.

There was one tiny plus. Dredging in the West Basin virtually had been eliminated by the sinking of land. In the early days of the subsidence problem, the statement had been made publicly and not contested that the money spent for subsidence remedial work in the year had not matched that spent at other yards for dredging alone. As the cost of repairing subsidence-damaged facilities mounted, the statement was no longer given currency in support of the continued operation of the Navy yard.

Navy, of course, was not the only one affected by disappearing land. Going down with Navy in the subsidence bowl was a conglomeration of factories, shipyards, docks and warehouses. Almost in the center of the pit, Southern California Edison Company operated a $22 million steam plant. The company had spent more than $7 million trying to save the plant, which lay 15 feet below high tide. Ford Motor Company's assembly plant a mile away lay below sea level most of the time. Ford had spent $4 million to protect its facility. Other harbor industries in jeopardy included Proctor & Gamble, Craig Shipbuilding, Spencer-Kellogg and Kaiser Gypsum, who among them employed more than 1,000 people representing an annual payroll of more than $6 million.

The City of Long Beach was not immune. Streets had buckled, sewers and underground pipes had broken, the towers of the Commander Heim Bridge had tilted, store windows in downtown Long Beach had cracked. The city hall had moved three feet closer to the center of the bowl, and the famed Rainbow Pier and Municipal Auditorium were threatened.

As if all that were not enough, there was something else of tremendous value at stake. It was the future of California's largest oil field.

When one drew the outline of the area that was sinking, the center of the egg-shaped subsidence bowl coincided with the heart of the Wilmington field. The field not only was California's number one field but also the second largest that had ever been discovered in the United States, runnerup only to the huge East Texas field.

Even as Southern California Edison Company tunneled beneath its steam plant to shore up the foundation, those with oil wells fought a desperate battle to preserve productive capability. The sinking of the land's surface was only part of their problem. It quickly became apparent there was another force at work, a deadly unseen force that threatened oil wells far below the surface of the ground.

Production unaccountably began to fall off at some wells. Others stopped producing altogether. When operators went in to see what remedial action might be required, they found casing crushed in some wells, completely sheared off in others. The damage seemed to be occurring at depth of about 1,600 feet. It was as if an unseen vise, or huge hand, had gripped the casing, choking it until the casing could no longer protect the well through which oil was produced.

An early survey showed some 50 wells to be affected, including 22 operated by the City of Long Beach and others

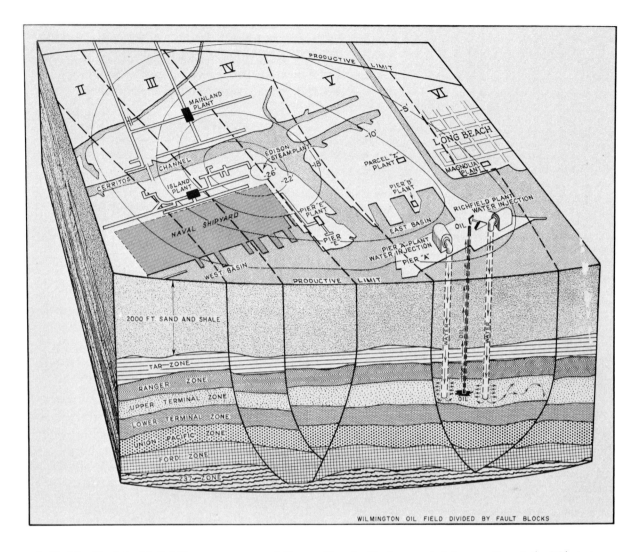

WILMINGTON OIL FIELD DIVIDED BY FAULT BLOCKS

drilled by Richfield Oil Corporation and Union Pacific Railroad. Adjacent wells were suspected of having minor damage. The wells that were having trouble seemed, by coincidence, to be those in the center of the subsidence bowl in an area extending northward from tidewater to a point north of Anaheim Road and westward from the Los Angeles Flood Control Channel to a point several hundred feet west of the Ford plant.

Artist's conception of the Wilmington field illustrated the six fault blocks and the field's seven producing zones. Surface contours indicated the extent of subsidence (Port of Long Beach)

Operators managed to get some relief by rolling out bent casing and squeezing cement through perforations, but it seemed questionable whether such measure could withstand the apparent lateral movement of a ten-foot interval activated by some unknown force.

The survey of wells was extended to include some 160 wells in the area most affected by subsidence. Concerned operators found deformed casing in 96 of the wells. The damage was so severe in 22 wells that operators had no choice but to abandon them, representing a loss of about $2 million, not to mention the value of the oil that might have been recovered. For other wells with damaged casing, it looked like it would be only a matter of time before they, too, would have to be abandoned. Many of the wells produced substantial volumes of a powdery sand whch periodically had to be removed to permit effective pumping. It seemed obvious that when such wells sanded up and it became necessary to do remedial work, the operators stood a good chance of losing the wells.

The situation at Wilmington was serious, not only for the operators whose wells were immediately involved but also for the future of the field. Development had begun at Wilmington in 1936 when General Petroleum Corporation brought in Terminal No. 1 on Sec. 4, 5S-13W, Los Angeles County, flowing 1,389 barrels a day of 20.5-gravity oil from a depth of about 3,000 feet. Since then various operators had completed more than 2,400 wells through which they had produced more than 800 million barrels of oil. By most estimates, the amount of oil that could still be produced from the developed area of the field amounted to more than 200 million barrels.

The value of the reserves in the developed portion was small compared with what was believed to underlie the undeveloped portion of the field that extended eastward beneath the tidelands. Most geologists were confident that at least one billion barrels of oil lay beneath the undrilled seaward extension of the field.

The untapped tidelands oil was a wildcatter's dream. The huge reserves were located within 10 miles of a number of major refineries in the nation's largest gasoline-consuming area. A large portion of the indicated field was in a breakwater-sheltered offshore region where water depths rarely exceeded 50 feet. The structure had been thoroughly mapped geophysically and all indications were that it would be highly productive. But fear of subsidence hung over Long Beach, and there was reason to believe the seaward extension might never be developed.

As quickly as subsidence was perceived as a threat, going back into post-war days of the 1940s, the Long Beach Board of Harbor Commissioners, which administered the tideland oil properties, Navy and others with interests in the affected area engaged engineers, geologists and other scientists to study the strange sinking to determine what was causing it and how far it might go.

Almost from the beginning, U.S. Grant III, geology professor at the University of California at Los Angeles, and others theorized that subsidence and oil production were related. Grant, in a joint paper with Geologist James Gilluly published in the *Bulletin of the Geological Society of America* in 1949, concluded that subsidence at Wilmington was the result of oil field operations. Grant and Gilluly found close

This transit shed on the northern end of Pier A, Berths 1 and 2, was being raised because of subsidence. The ground surface had sunk so far that ocean water occasionally washed over the shed at high tide. The transit shed was moved back to the new elevation. (The Port of Long Beach)

agreement between the relative subsidence of the various parts of the Wilmington field and the pressure decline due to oil production, thickness of the oil sands involved and the mechanical properties of the sands.

For many, the most convincing indictment of oil production as the cause of subsidence was an ingenious bit of downhole sleuthing performed by Jan Law, a Harbor Department consultant. Law, working with Harbor Department petroleum engineers and an oil field service company, devised a magnetic "collar-counter," which gave indication electrically when it came opposite a collar, that is, the connection between joints of pipe, in the hole.

When the collar-counter was run on a wire line, engineers were able to measure pipe length far down in the well. Since each piece of pipe had been measured before it went into the hole, a comparison of the pipe's length with the original measurement showed whether there had been any compression lengthwise.

Law's research demonstrated that the casing in wells was being telescoped by compaction in the Wilmington field's four major oil-producing zones. The zones and the average depth at which each was productive were the Tar,

Sinking ground left wellheads stranded high and dry while worried operators trucked in fill to build up protective dikes. (The Port of Long Beach)

2,200 feet; Ranger, 2,500 feet; Upper Terminal, 3,000 feet; and Lower Terminal, 3,500 feet. Some 40-foot joints of casing had been shortened as much as one foot.

Soon afterward, G.D. McCann, professor of electrical engineering, and C.H. Wilts, assistant professor of applied mechanics at California Institute of Technology, came up with another study that suggested an answer to the question of what mysterious force was causing the crushing and shearing of casing at a depth of about 1,600 feet in some wells near the center of the subsidence bowl.

They concluded that the continued subsidence gradually set up horizontal shear stresses that finally overcame the shear strength of the material in the planes of slippage, resulting in damaging earth movement.

McCann and Wilts found a direct correlation between actual subsidence and a data factor resulting from multiplying the drop in pressure in the oil reservoir by total thickness of the productive oil sand. They derived a mathematical formula and used it to calculate theoretical subsidence for various points throughout the bowl for the years 1947 to 1951. They used the data to plot curves and when they compared the curves with those determined by actual measurement, the two showed remarkable agreement.

Land subsidence was not a new phenomenon. It had occurred in various parts of the world. In California, it had been observed since 1922 in the peat lands of the Sacramento-San Joaquin Delta region and since 1930 in various other areas of the San Joaquin Valley, amounting at some places in the valley to as much as 10 feet. The subsidence had been publicized due to the menace to the Delta-Mendota and the Friant-Kern Canals, but for the most part, sinking had occurred in places where surface consequences did not seem serious.

Before Wilmington, there had been many large oil fields developed in or near cities, and in none of them had subsidence created a problem. If it had occurred at all, it had not been of sufficient severity to result in damage to surface structures. No one had heard of subsidence in connection with the tremendous oil developments at Signal Hill, Santa Fe Springs or Huntington Beach. One of the great oil fields of the United States lay in the heart of Oklahoma City. Subsidence had not occurred there. There had been no previous experience to indicate that oil production might cause land surface subsidence. Why, then, had it occurred at Wilmington, where the problem was complicated by the fact that the affected area of more than 10 square miles was carrying with it into a cavity already 21½ feet below the old surface of the ground the industrial complex of a large city?

Among those making studies at Wilmington was Richfield Oil Corporation, which operated oil wells in the field and also had valuable marine terminal facilities near the center of the subsidence bowl. The company's study concluded that three characteristics of the Wilmington field's productive pools seemed to explain why the production of oil had resulted in the unusual subsidence of the land.

First, unlike other oil reservoirs in California, which were strongly folded, with dips of up to 45 degrees on the flanks, making a strong arch, the Wilmington field's reservoirs were relatively flat, with dips of only 10 to 15 degrees on the flanks.

Second, Wilmington's reservoirs were relatively close to the surface, and instead of being overlain by thick layers of shale which occurred in most oil fields, they were overlain by thin layers of shale and by gravel and sand.

And finally, the Wilmington reservoirs were extremely thick, but at the same time they were not well cemented or consolidated. When the reservoir pressure was reduced, there was a greater than usual volume of material subject to compaction.

In a speech before the Rotary Club of Long Beach in August, 1955, Charles S. Jones, Richfield president, forecast a grim future if nothing was done. Jones said the company's studies indicated that if production continued from the present developed area and if lands belonging to the City of Long Beach in the offshore portion of the field were never developed, subsidence eventually would be 35

feet at the Southern California Edison plant—or some 14 feet more than it already was, 5 feet at the Municipal Auditorium, 4 feet at the Villa Rivieria and 3 feet at the intersection of American and Anaheim Streets.

If the seaward extension of the field were to be developed, Jones said, the studies indicated subsidence would be no deeper at the Edison plant but would be 17 feet at the Municipal Auditorium, 12 feet at Villa Rivieria and 4½ feet at the American and Anaheim intersection. He added that subsidence also would seriously affect such new developments as Belmont Shore, Naples and Alamitos Bay.

What could be done? What were the possible remedies? The first thing that came to mind was that if the production of oil was causing land to sink, then why not stopping producing oil? Why couldn't the city of Long Beach set an example, shutting in wells on city property?

The Wilmington field was being developed on both east and west sides on lands not belonging to the city and outside the city limits. If the city stopped producing its properties, it would merely lose oil to others through

Subsidence caused the northern portion of Pier A in Long Beach Harbor to sink so far that at times ocean water would wash over the wharf. (The Port of Long Beach)

drainage by adjoining properties. Subsidence would slow, but it would still continue. And if the city were to attempt to stop production through the entire field, it would require state legislation and the result would almost surely be endless litigation.

Moreover, it seemed unthinkable to stop the production of oil as long as there was any other possible remedy. When one spoke of stopping production at Wilmington, one was talking about placing in limbo more than one billion barrels of oil, and oil semed far too vital to national security and too important to the economy of California even to consider leaving that much in the ground, never to be produced.

A second possible remedy was to continue doing, but to an even greater degree, the things that had been done so far—the construction of dikes, the raising of foundations, the filling in of land. This would save property, but it would not save property values, because it would not save the beaches and it would not save Long Beach as a recreational and tourist center. Furthermore, it seemed foolish to spend money to build dikes, raise foundations and fill land, because none of those measures would really get at the root of the problem

To meet the problem of subsidence, Jones and others suggested a way to attack the cause. Contrary to stopping the production of oil, it would result in the production of much more oil. Contrary to spending money to build dikes and take other remedial measures, it would result in far greater revenue to the city, and over a longer period of time.

The suggested remedy seemed disarmingly simple. If subsidence was caused by the reduction of reservoir pressure due to the withdrawal of gas, oil and water from the reservoir, then why not stop further subsidence by adding to the reservoir an equivalent volume of water? By injecting water under pressure in back of the oil-water contact, great pressure would be added to the water and it would act like a giant piston pushing oil ahead of it to the areas of low pressure where the oil could escape through wells.

Waterflooding, discovered accidentally in Pennsylvania in the 1880s, had been practiced for many years in other states as a means of increasing production. Until recently, however, it had been thought that flooding would not work in California fields because reservoirs tended to be small in area, very thick and more faulted.

Flooding on a pilot scale had not begun in California until 1944. Since then Richfield Oil Corporation, Shell Oil Company, Standard Oil Company of California and Union Oil Company of California had all reported some success with waterfloods. In the Wilmington field itself, pilot floods carried out by Long Beach Oil Development Company and Union Pacific Railroad had indicated waterflooding would increase recovery and restore formation pressures. LBOD's pilot had produced 1.4 million barrels of oil not recoverable through primary means and restored pressures from 930 pounds per square inch to the original 1,200 pounds per square inch. Union Pacific's pilot had raised pressures from a low of 500 pounds per square inch to the original 1,100 pounds per square inch and produced an estimated additional 300,000 barrels of oil.

The Pier E water injection plant immediately east of the Long Beach Naval Shipyard brought to bear a capability of 135,000 barrels-per-day water injection to boost the battle to halt subsidence in the shipyard area. (The Port of Long Beach)

If waterflooding could not only offer hope of arresting subsidence but also add millions of barrels to ultimate recovery from the field, why had it not been started long ago?

Merely to state the proposition was to explain the impossibility of its accomplishment. For waterflooding to succeed on a fieldwide scale, it would have to involve the operation of each of the oil and gas reservoirs in the field as a unit. Under unit operation, the production from the entire reservoir would have to be divided on a fair basis between all of the separately owned properties included in the reservoir, but property lines would have to be completely disregarded so far as operations were concerned.

Some wells would be selected for water injection, others for oil production at strategic places with relation to the geometry of the reservoir but with no relation whatever to property lines. Rates of production would have to be carefully controlled and might vary at different places in the reservoir, and such control also would involve complete disregard of property lines. All wells on one property might be closed down entirely as producing oil wells and be used only for the injection of water and the pushing of oil from the property to one more favorably located for greater ultimate production of oil.

Unit operation would mean a single control of the reservoir energy, as that would be the only way possible to obtain the greatest ultimate production from the reservoir. It would be the only way in which it would be possible to protect the right of each property owner to his fair share of the oil.

In the Wilmington field, there were thousands of different surface owners, and more than 100 different operating interests. Under existing law, it would not be possible to have unit operation unless there was an agreement to which 100 percent of all of the different owners were parties.

It was obvious that legislative help would be required. The City of Long Beach with the support of its citizens sought state assistance in enacting special legislation.

Water injection well adjacent to the Long Beach Naval Shipyard was drilled in an effort to halt subsidence. This well, and dozens like it, permitted salt water from high-pressure pumping plants to be pumped underground to bolster subsurface forces. (The Port of Long Beach)

Faced with the threat of inundation of the Long Beach area and with the closing of the Navy shipyard, the Legislature's Assembly Interim Committee on Manufacturing, Oil & Mining Industry held two public hearings late in 1957, including one in Long Beach and another in Los Angeles.

From the hearings and study of the testimony the Committee concluded that the only feasible method to arrest or reduce subsidence was by repressuring the oil zones underlying the subsidence area; that for efficient repressurization on an equitable basis, cooperative operation among the owners of oil properties would be necessary; and that because of the large number of people affected, including about 130 operators and several thousand interest owners, compulsory utilization might be necessary.

On March 4, 1958, a Charter election was called in Long Beach to seek an amendment to the Charter permitting the city to enter into unitization agreements. Voters favored the amendment 15 to 1.

Later in the same month, Governor Goodwin J. Knight called a special session of the Legislature to consider the subsidence problem. The Legislature quickly enacted the Subsidence Control Act of 1958, effective July 24. The Act encouraged voluntary pooling and unitization but provided for compulsory unitization if necessary. It provided rules for determining costs of initiating and conducting a repressuring operation and set up procedures for filing of plans of operation. It also provided for hearings and for appeal or judicial review of cases where interested owners objected to the formation of a unit. In the event that voluntary programs might fail, State Oil & Gas Supervisor E.H. Musser was empowered to order unitization if the order was acceptable to working-interest owners entitled to 65 percent of the proceeds of production from the proposed unit. If ratification of the order could be achieved from persons entitled to 75 percent of the proceeds of production in the proposed unit area, a city or county might exercise the right of eminent domain to acquire properties of nonconsenting persons.

The federal government made it plain it would not tolerate any delaying action. The government on August 15, 1958, filed in the U.S. District Court a $54 million property damage suit and an injunction to compel the City of Long Beach, the State of California and oil operators in the Wilmington field to take immediate steps to halt subsidence in the Long Beach Naval Shipyard and other government installations in the area.

The stage was set for an all-out war on subsidence. To determine the boundaries of the area to which the Subsidence Control Act would be applied, State Oil & Gas Supervisor Musser held public hearings in September in Los Angeles. One of those offering testimony was Mayor Raymond C. Kealer of the City of Long Beach, who warned that the situation had developed to where it was endangering human life. He asked that special attention be paid to the southeastern section of the field in the Alamitos

The island water injection plant, so-called because of its location on Terminal Island immediately north of the Long Beach Naval Shipyard, pumped 157,000 barrels a day of salt water into the ground for repressuring of oil sands beneath the subsidence bowl. (The Port of Long Beach).

Bay, Belmont Shore and Naples area, stating that the communities were built for the most part on filled land, and if they sank any more, they would be in danger of inundaton from any unusual tides or especially inclement weather. "We can't sit back and wait for these tragedies to happen," the Mayor said.

In the wake of the hearings, the State Oil & Gas Supervisor established exterior boundaries of lands comprising the subsidence area at or near the limits of the Wilmington field, including the eastern extension into the tidelands. The subsidence area comprised approximately 21,600 acres of land and tidelands, including all of Long Beach Harbor and the eastern portion of Los Angeles Harbor, extending to the southeast as far as the Orange County line, to the north to Lomita Boulevard and to the west as far as Figueroa Street.

The problem of waterflooding the Wilmington field was complicated by the fact that the field was a complex one, divided into six major areas by faults. The faults had caused huge subsurface areas known as fault blocks to be displaced, with each block so separated from the others that water would have to be injected into each fault block individually.

To add to the complexity, there were seven major oil-bearing sands, separated from each other by layers of impervious shale. Within each of the major sand intervals were several minor sand bodies separated by thin layers of impervious shale. Because of the individual characteristics of the sand bodies, each would require a different amount of water to increase the pressure properly. Consideration would also have to be given to the fact that the injected water would move oil ahead of it. A poorly engineered program could result in the loss of millions of dollars worth of oil.

To accelerate the control of subsidence in critical areas, six fault block unitization areas were formed rather than one fieldwide unit. The activities of each fault block were placed under the direction of a management panel containing representatives of the participating parties. The

detailed engineering, accounting and legal work was to be performed by six technical committees, that would provide information for the hundreds of decisions that would have to be made.

The top priority was given to the Long Beach Naval Shipyard and the area underlying downtown Long Beach. Pressure restoration in the area beneath the downtown district was assured through agreements concluded in February, 1959, between the city and private operators. Producers in the vicinity of the shipyard did not wait for final unitization agreements before beginning water injection. They commenced flooding under terms of less complicated agreements.

The Wilmington field quickly became the busiest oil field in California from the standpoint of drilling activity. Eight rigs went to work, including four for the City of Long Beach, three for Richfield Oil Corporation and one for

The city of Long Beach-owned Mainland injection plant north of Cerritos Channel in Long Beach Harbor was the largest of its kind in the world. Built at a cost of $1,208,000, it brought a capacity of 287,000 barrels of salt water per day to the Wilmington waterflood. (The Port of Long Beach)

Long Beach Oil Development. Instead of oil wells, they drilled water injection wells to bolster what soon would become the world's biggest waterflood. Contractors furnishing rigs for the program were Camay Drilling Company, four rigs; Fred Johnson Drilling Company, three rigs; and F.E. Gober Drilling Company, one rig.

The City of Long Beach undertook the responsibility of constructing the necessary water plants and supply systems, spending some $6 million in the work to further encourage private parties to participate by lowering their initial capital outlay cost. The city would recover its investment through the sale of injection water.

On the afternoon of October 29, 1959, the Long Beach Harbor Department hosted a half-day event to celebrate the greatest stride that had yet been made in the development of the world's largest waterflooding project—the dedication of the Terminal Island and Mainland water injection plants. On hand were some 200 people, including government officials from city, state and national levels, newspaper reporters from such distant papers as the *London Times,* the trade press and television personalities.

At a given moment, Captain C.J. Palmer, Commander of the Long Beach Naval Shipyard, turned a valve in the field, and water arched into the air from a standpipe 20 feet to the rear of where the Captain stood. The pipe had been placed so neither the water nor its spray would touch those on hand for the ceremony.

This was "Big Squirt," a dramatic portrayal for the media of the stream of water that would be pumped into the ground to stop the sinking of the City of Long Beach, its people and its multimillion-dollar industrial complex.

The water gleamed in the afternoon air. As it mounted hundreds of feet into the sky, it formed a spray, and the fragment of a rainbow appeared, nature's good omen.

The plants that were being activated had a vast network of over 20 miles of water distribution lines with a planned capacity of nearly one-half million barrels of water per day to serve the two fault blocks which covered the critical Navy shipyard area.

M.A. Niskian, head of the Long Beach firm that had designed the $1.9 million plants and their water distribution systems, explained that the purpose of the systems was to take salt water from source wells and distribute it to injection wells, where the water would be injected into the subsurface at high pressures ranging from 850 to 1,250 pounds per square inch. The reason for taking salt water from source wells instead of the ocean was that the natural filtration of the water by the earth and the water's low oxygen content made its use preferable.

In time, those present for the dedication boarded buses for the trip back to briefing headquarters in the Long Beach Petroleum Club for cocktails and a steak dinner. As buses pulled out, some of those aboard looked back to see where the still cascading stream of water might be landing. The water was raining down with cloudburst intensity on a boxcar-like structure that was the doghouse for the crew of a well-pulling rig and on the little out-building that was the crew's chemical toilet.

Five months later, U.S. Attorney Laughlin Waters announced that in view of the satisfactory progress being made in the $30 million Wilmington waterflood's control of subsidence in the shipyard area, the government's injunction suit against oil operators would be kept off the court calendar.

In 1964, the Wilmington field produced its one billionth barrel of oil, becoming the first oil field in California to reach that mark and the second in the nation, following the lead of the East Texas field.

Even as production was reaching that milestone, operators were injecting half a million barrels of water a day into the field, or more water than the field's average production of 97,000 barrels per day of oil.

Surveys showed that pressures in the upper four zones of Fault Blocks II through VI—those most seriously involved in the subsidence problem—had increased since 1959 by as much as 550 pounds per square inch. It was estimated that the amount of oil produced over and above the field's estimated primary recovery was 60 million barrels and still climbing, and the volume of water that had been injected was in excess of one billion barrels.

No subsidence was noted in downtown Long Beach, in areas overlying Fault Block VI or on the north and south flanks of the structure in Fault Blocks IV and V. Elsewhere, there seemed to be a definite decrease of land sinkage. Subsidence at the center of the bowl was slightly more than 28 feet.

In its annual report for 1968, the state's Division of Oil & Gas said the entire Wilmington-Long Beach area seemed to be undergoing a regional-type uplift, with the maximum rebound being four inches. The bench mark on Terminal Island that marked the point of maximum subsidence gained one inch in elevation during the year. It marked the first time, the Division reported, that the cumulative subsidence figure had reversed its trend.

7 The Mighty Mole

Usually when a drilling contractor wanted to introduce a new drilling rig, he invited some friends and potential customers out to the yard where the outfit had been rigged up, led them up and down steel steps pointing out the various features of the power package, mud system and control console and afterward set up drinks at a makeshift bar and put on a barbecue with all the trimmings including beans, tossed green salad, French rolls and plenty of hot black coffee.

It was a departure when one contractor in 1958 chose to announce a rig showing with an artist-designed, engraved invitation which read:

"Brown Drilling Company cordially invites you to see The Mighty Mole ... the latest and finest in dual trailer-mounted drilling equipment.

"On Thursday and Friday, April Ten and Eleven from 10 A.M. until 4 P.M. at the yard of Bender Engineering and Manufacturing Company, 1620 South Union Avenue, Bakersfield, California."

On the invitation was an insignia that depicted a mole so engaging that the burrowing insectivore might have stepped intact from the latest Walt Disney animated feature. The cartoonist's mole smiled confidently while flexing powerful biceps. On the mole's head an oilworker's hardhat rode at a properly jaunty angle. The cartoonist had chosen to put a splash of bright orange color on the mole's sturdy chest and another dab on the animal's nose, which set off the insignia with just the right touch of eye-catching appeal.

Brown Drilling Company had decided in a major policy decision just executed by management to retain the services of the Jakobsen Agency, a Los Angeles public relations and advertising agency, to publicize and advertise what a Brown official described as the contractor's "progress, accomplishments, personnel and development."

There was no mistaking the Hollywood touch in the Jakobsen Agency's introduction of Brown Drilling Company's newest drilling rig to the California oil industry. (Brown Drilling Company)

The drilling company, whose headquarters were in Long Beach, had made its debut in 1937 as one of California's first drilling contractors.

Along with setting up an advertising campaign for the Mighty Mole, the Jakobsen Agency had commissioned a famous artist, Paul Landacre, to make a series of wood engravings to be used to depict the romance and glamour of the oil drilling business while calling attention to the availability of Brown Drilling Company rigs for domestic and foreign assignments.

In addition to the Mighty Mole that was about to join its list, the drilling company offered five other rigs in California, including four that worked out of an office and yard at Willows in the Sacramento Valley and another on Monterey island in Belmont Offshore field. Overseas, the company was running four rigs in Venezuela, two in Turkey and two in France. Brown rigs also had worked in Saudi Arabia, Bahrain Island, Australia, New Zealand and on the island of Trinidad.

In *Diggin' Dope,* a Brown Drilling quarterly published in tabloid newspaper form as an in-house organ, a company spokesman acknowledged that it might be hard for the unimaginative to see great art in the cold steel of an oil drilling operation but added that Paul Landacre saw it and, using Brown rigs as an inspiration, had begun a series of pictures that were to be widely distributed as a tribute to the colorful side of the oil industry.

The artist was a winner of many awards for his art achievements, had illustrated numerous books and was on the faculty of the County Art Institute at Los Angeles and also a member of the National Academy of Design. His prints had been purchased by innumerable galleries and libraries, including the Museum of Art, the Library of Congress, the New York Public Library, the Victoria and Albert Museum in London, the Art Institute of Chicago and dozens of West Coast institutions.

While Landacre spent nearly a week making preliminary sketches at a Brown rig that was working near Willows, Bob Jakobsen of the Jakobsen Agency carefully planned the

The Mighty Mole represented a half-million dollar approach to the challenge of drilling wells faster and cheaper. (William Rintoul)

Bakersfield showing at which Brown's innovative new drilling rig would be introduced. Jakobsen even made arrangements for hostesses. They would be drawn from a pool of Western Airlines stewardesses who moonlighted at such functions.

The wonder was not so much that Brown had turned over the job of showing the rig to an advertising agency but that the drilling contractor had chosen to build the rig at all. The times hardly seemed propitious for Brown or anyone else in California to spend money building a rig, no matter how innovative the iron might be.

The drilling business had fallen on hard times. A rig count showed that only 86 of 214 contract rigs available for assignments in California had jobs, i.e., 40.2 percent working. In a period of three months, six drilling contractors had gone out of business. The contractors had operated a total of 19 rigs.

The drilling slump was a reflection of serious problems. Production was plummeting in California, inventories

On hand to show visitors around the Mighty Mole at the Bakersfield showing were, left to right, Ned Brown, chairman of the board and founder of Brown Drilling Company; Bob Sharp, vice president, Brown International; and Art Heiser, vice president and soon to be president of Brown Drilling. (William Rintoul)

were mounting and the purchasers of crude were cutting prices.

The decline in production had set in immediately after the peak of slightly more than one million barrels daily in 1953. By 1958, output had dropped to 865,400 barrels a day, representing a decline of some 135,000 barrels daily.

There was more to the grim picture than the natural decline one might expect from maturing fields, or the failure of wildcatters to find new fields.

As Brown Drilling Company put the finishing touches on the Mighty Mole, some 60,000 barrels a day of California production stood shut in—distressed oil without a buyer.

The situation, instead of showing any sign of improvement, was getting worse. Inventories were mounting, building up pressure for crude price cuts that before the end of the year would see a reduction in posted prices of from 10 cents a barrel for high gravity oil to 50 cents a barrel for heavy crude.

To California's independent producers, there was no mystery about what had caused the downturn. The problem was not that consumers had suddenly quit using gasoline or other petroleum products. Demand had continued to rise. The truth was that California oil could not compete with foreign oil.

The trickle of imported oil that had begun when California was no longer able to supply its own needs had swollen until foreign oil not only made up the difference but also started to drive domestic oil off the market.

It was a simple matter of economics. Thanks to a worldwide surplus and extremely low tanker rates, a barrel of 34-gravity crude from the Middle East could be laid down at a Los Angeles refinery for $2.95. That was 65¢ a barrel less than the $3.60 that it cost to lay down at the same refinery a barrel of like-gravity crude produced in the Los Angeles Basin.

The crude produced in Venezuela and Canada also enjoyed a competitive advantage. A barrel of Venezuelan crude could be laid down at a Los Angeles refinery for

$3.40, or 20¢ a barrel cheaper than oil produced in the Los Angeles Basin. A barrel of Canadian crude could be laid down for $3.55, or 5¢ a barrel cheaper.

By 1957, the flow of imported crude had increased to an average of some 250,000 barrels a day. Independents from California and from other oil-producing states that had been hard hit by imports went to Washington, seeking protection from the rising tide, arguing that there was no security in foreign oil.

President Eisenhower in July, 1957, responded with a program calling for "voluntary" curbs on imports. Though West Coast importers were exempted from voluntary curtailment on grounds the district was not able to supply its needs, importers still cut back, so that by the spring of 1958 the floodtide of foreign oil was reduced to about 183,000 barrels a day, with some 50 percent coming from the Middle East, principally from Saudi Arabia and Kuwait, about 20 percent each from Venezuela and Sumatra and some 10 percent from Canada. The importing companies included General Petroleum Corporation, Shell Oil Company, Standard Oil Company of California, The Texas Company and Tidewater Oil Company.

California's woes continued to mount. Stark Fox, executive vice president of the Oil Producers Agency of California, flew to Washington to appeal to Captain Matthew V. Carson Jr., administrator of the government's voluntary imports program, and to California Senators William F. Knowland and Thomas H. Kuchel for help in further reducing the flow of foreign crude.

Even as Fox was asking for a cutback in imports, Otis H. Ellis, counsel for the National Oil Jobbers Council, was appearing before the house Ways and Means Committee to speak against a proposal that a law be passed to control oil imports. The Jobbers Council opposed any cut in imports of cheap foreign oil on grounds that such a cutback would lead to higher prices for consumers in the heavily populated, industrialized northeastern part of the United States. Ellis warned that if a quota or tariff system were legislatively imposed, "Complete federal control of the

domestic petroleum industry, in all its aspects, will come."
He told the Committee that the independent producer
should be "willing to take a few jolts without running to
Congress every time he stubs his toe."

Another witness who spoke up strongly against a man-
datory controls program was T.S. Petersen, president of
Standard Oil Company of California, which was bringing
some 28,000 barrels a day of crude into California from
Sumatra. Petersen termed the world oil surplus "transi-
tory" and added, "Were we to bar much of the flow of this
oil from our shores, with a reversion to the foreign trade
philosophy of the McKinley era, our friends and allies
abroad would have good reason indeed to conclude that
our intentions toward them were something less than
honorable.

"What other alternative would they have than to deduce
from this that any other important commodity they had
for export could be and would be similarly barred, whe-
never it suited our convenience?"

Foreign oil was not the only factor depressing the California production picture. From supplying slightly over 20
percent of the total energy demand on the West Coast in
1950 natural gas had grown to where by 1958 it accounted
for nearly 35 percent of the energy market. Two-thirds of
the gas used in California came from out-of-state, delivered by pipeline from Four Corners gas fields and from
Texas.

The price of gas was federally regulated, which made it
possible for Los Angeles Basin utilities to purchase out-of-
state gas for electric power generation at about 35 cents
per thousand cubic feet. When converted to the energy
equivalent of a barrel of fuel oil, this worked out to be less
than $2.10 a barrel, compared with fuel oil's posted price of
$2.45 per barrel.

Against this background of slumping production, shut-
in wells and slack drilling activity, Brown Drilling Company offered the Mighty Mole to the California oil
industry. Ned Brown, the firm's founder, who had
recently stepped up to the position of chairman of the

Though light in weight, the Mighty Mole carried a capability for drilling to 9,000 feet. (William Rintoul)

board, and Art Heiser, who had taken over as vice president and before the year was out would be named president, showed interested visitors around the rig.

Brown was known throughout the drilling industry as an innovator. Some 20 years before, he had pioneered in the Sacramento Valley with what was believed to be the first portable drilling mast ever used in the state, putting the equipment to work on a job near Rio Vista.

Heiser, a graduate of Colorado School of Mines, had worked for Brown Drilling early in his career as a petroleum engineer, then in partnership with Don Green had formed Green & Heiser Drilling Company. The two had made a down payment of $27,000, their combined life savings, on a $268,000 drilling rig and parlayed the investment into a two-rig operation. Green had died in an airplane crash in Oregon, and Heiser had sold the rigs to Brown Drilling and returned to the latter company as vice president.

"I think it is noteworthy," Heiser said at the Bakersfield showing, "that at a time when half the rigs in California are idle, Brown Drilling Company is showing its faith in the future—and in the oil industry generally—by pioneering this cost-cutting equipment."

The Mighty Mole represented a bold approach on the part of Brown Drilling in concert with the design, engineering and manufacturing capabilities of Bender Engineering & Manufacturing Company. It was a half-million-dollar rig designed for fast, easy moves and quick rig-up even in the most difficult terrain.

About half the Mighty Mole's structure consisted of lightweight metals, including large amounts of special aluminum, high-strength steel alloy and manganese steel. It was anticipated that the light weight would conserve hauling costs and also speed up the placement of equipment on jobs in difficult terrain.

Though the rig was light, it had a powerful drilling punch. Five big diesel engines provided a total of 1,500 horsepower, giving the rig a depth capability of 9,000 feet.

The rig's extreme portability was attained through use of dual trailers, which could be easily positioned by a mating arrangement with guide trucks and fastened with patented wedge lock connectors. Only one connection had to be broken to separate the drive between the mast trailer and the drawworks trailer.

The rig moved in 13 compact loads with 8,000 feet of 4-inch drill tubing. All of the loads were road legal except for the mast load, which was permit legal.

There were such special features as unitized and prefabricated surface mats especially prepared to permit quick leveling and easy alignment of the mast trailer with the drawworks trailer, and aluminum steps which two men could put in place without any assistance from a crane.

The principal components included a Bender 126-foot guyless and telescoping mast, an Emsco GC-350-T double drum drawworks powered by two GMC turbocharged diesels rated at 348 horsepower each at 1,800 revolutions per minute; two skid-mounted Emsco mud pumps including a D-300 driven by one supercharged GMC diesel and a DA-500 driven by two supercharged GMC diesels; a 196-barrel skid-mounted Bender mud-shaker tank; another mud-suction tank of the same specifications; and a portable combination water, mud and fuel tank 40 feet long and 8 feet in height and width, which included a 2-horsepower circulating pump and storage for 300 barrels of mud, 200 barrels of water and 3,000 gallons of fuel.

Shell Oil Company's package drilling rig was designed to cut the time and cost of drilling deep wells in the Ventura field. (William Rintoul)

The wheels of the combination tank could be retracted so that the tank formed an anchor for mastload lines and the drawworks trailer.

Following the Bakersfield showing, the Mighty Mole left the Bakersfield yard for a day's travel northward to Roberts Island, six miles southwest of Stockton. On location, overall rig-up time was only 7½ hours, including 4¼ hours for crane time. The Mighty Mole spudded in on

Gene Reid Drilling Inc.'s Hopper Slimhole rig made its debut on assignment for Golden Bear Oil Company at New Hope No. 13, a 2,700-foot test of Pliocene sand in the Poso Creek field. (William Rintoul)

April 24 to drill Standard Oil Company of California's Capital Company No. 2-1, a gas test. The rig completed the hole 12 days later at a depth of 5,358 feet and moved to the Wood Community lease to begin another well for the same company.

The fact that anyone would dare to put together a costly new rig in the face of a declining market might have seemed foolhardy to some in other industries, but for the oil industry, it was the manifestation of a recurring theme that dated to the beginnings of the industry in Pennsylvania one hundred years before: the drive to build a rig that would drill holes faster, better and cheaper.

The Mighty Mole, as might have been expected, was not the only innovative rig to make its debut in California in the problem-plagued 1950s.

At Ventura, Shell Oil Company in cooperation with the Ideco Division of Dresser Equipment Company came up with a package drilling rig designed to cut time and cost of drilling wells to depths of 10,000 feet and more.

Engineers designed the rig with a minimum number of packages, which facilitated tearing out and rigging up operations and also, since there were fewer packages to transport, shortened the time of moving between locations.

The rig featured an Ideco Full-View mast and Ideco 1350-S drawworks and was the largest torque-converter rig with 100 percent friction-clutch operation placed in service up to that time.

The draw works and special torque-converter transmission were designed to take advantage of the characteristics of torque converters and air clutches to permit clutch engagement to be made prior to advancement of the throttle. This made it possible to accelerate the load with minimum wear and shock on clutches, chains, bearings and shafts throughout the entire hoist, transmission and compound. There were three forward speeds and one reverse speed in the transmission and all speed changes were made with air clutches. The clutch element was split and was replaceable in position without removing the shafts.

The Gene Reid rig represented a quarter-million-dollar investment designed for versatile service in drilling slimhole to depths of as much as 7,500 feet. (William Rintoul)

When the rig was moved, engines could be left in operating position or removed to break down the load to conform to eight-foot road-width requirements.

Grouping of controls in a console cabinet near the brake on the derrick floor enabled the driller to have fingertip control on virtually all the equipment used in the rig. From his position at the console, the driller had clear overall vision of the floor and derrick. A small doghouse with electric heat was supplied for the driller's comfort opposite the console.

For moving, the rig broke down into 11 loads, with each sized for handling on a heavy-duty truck or trailer. The mast, catworks, drawworks, transmission and mud-mixing unit moved in single loads. The compound units and pumps, mud tanks and two-piece box substructure moved in two loads each.

Three years before the Mighty Mole made its debut, Gene Reid Drilling Inc. had put together a rig in Bakersfield to offer versatile drilling service in the field of exploration. The rig was designed to drill slimhole, that is, 7⅝- to 8½-inch hole, to a depth of 7,500 feet. Within this hole it would be possible to run electric log, make formation tests and take sidewall samples.

With the slimhole rig, the operator would be able to evaluate thoroughly any and all horizons without having to drill the more expensive wider-diameter hole that would be necessary for a completion at depth. In short, the operator could find out if there was anything worth developing before he went to the expense of drilling a development well.

The rig featured a Hopper Hoistmobile as the basic hoisting unit. Incorporated with the Hopper Hoistmobile was an offset-bed, 40-foot semitrailer, which carried the rotary table, sand reel and cathead shaft. Power was furnished by a 6-71 GMC diesel and Allison converter-transmission combination.

A heavy-duty Kenworth Diesel truck tractor equipped with a special flush deck A-frame and retractable air-

A frothy crater marked the end of the trail for the Mighty Mole on Grizzly Island. (Gene Reid)

operated fifth wheel provided dual service both as a truck-tractor and A-frame.

The rig was entirely mobile with exception of three normal truck loads—the doghouse, shaker and mud pit, and pipe rack. Everything else moved under its own power under highway specifications. Gene Reid estimated that savings would amount to about a dollar a foot on transportation, rigging up and tearing out costs of wildcat wells.

The fact that no permits would be necessary when moving the rig from location to location represented real savings. The drilling contractor had found that three to five days were lost each month because of the necessity of obtaining permits from the Division of Highways for moves on Saturdays, Sundays, holidays and foggy or rainy days.

Hard times eventually idled the slimhole rig, and it was sold to Getty Oil Company. Crating specialists from Los Angeles boxed up the drilling outfit and it was shipped aboard the *Hoeghcliff* for a new base of operations in the Middle East's Neutral Zone.

Meanwhile the Mighty Mole kept reasonably busy in the Sacramento Valley gas country, drilling for various operators until in the spring of 1959, about a year after it had gone to work, it drew the assignment to drill a proposed 6,500-foot test for G.E. Kadane & Sons, a Newport Beach operator, on low-lying Grizzly Island in the Suisun Bay gas field 25 miles northeast of Oakland. The venture was the Standard-Suisun Community No. 10 on Sec. 31, 4N-1W, Solano County. The Brown Drilling Company crew spudded in to drill the well on June 2.

By the following day, the hole was at a depth of 1,024 feet. The daylight crew began pulling drill pipe from the hole to run surface casing. At 3:30 P.M. in the afternoon after four stands of drill pipe had been pulled, the well began to kick.

In a desperate effort to control the well, the crew weighted up the drilling mud from 72 to 135 pounds per cubic foot and using rig pumps supported by a Halliburton truck succeeded in getting away some 3,500 sacks of Baroid, a

The pipe rack was spared; the drawworks and other drilling equipment disappeared in the crater that was formed when the G.E. Kadane & Sons' well blew out. (Gene Reid)

weighting material, before 5:20 A.M., when the well started to crater. The crew had no choice but to abandon the rig. Fortunately, no one was injured.

As the substructure settled, the mast fell. The substructure, rotary table and blowout-prevention equipment disappeared into the boiling cauldron, which was estimated to be 25 feet in diameter. Gas, mud and water bubbled 15 feet into the air from the crater. The trailer-mounted drawworks and engines hung on the edge.

Brown Drilling Company quickly began putting timbers and gravel down to build a road strong enough to permit heavy tractors to close in on the threatened equipment and pull it to safety. With the water table near the surface, roadbuilders had their hands full trying to get the access route in.

The well blew wild with no apparent means at hand to kill it.

By the following day, the blowout appeared to be subsiding. Brown managed to pull the rig pumps to safety from the edge of the crater. The drawworks had settled three to four feet at the edge of the crater but was still visible, leading to hope that it might be salvaged. The rig's substructure, rotary table and blowout-prevention equipment were nowhere to be seen in the crater, and the mast lay badly damaged where it had toppled to the ground.

Oil men settled down to outwait the wild well, hoping the blowout would subside to permit cleanup and salvage work to begin.

Ten days after the blowout had begun, it continued unabated. A turbulent, muddy crater marked the site of the well, which was blowing gas at an estimated rate of three million cubic feet per day and flowing salt water at an estimated rate of 3,000 barrels per day. The water carried with it a fine silty sand and some blue shale, which stained the water a pronounced blue.

The Mighty Mole's mast, which had toppled soon after the blow began, still supported the drawworks, which teetered on the edge of the crater.

With the blowout showing no sign of abating, the operator began looking for another rig to drill a relief well to intercept the rampaging well at depth and enable crews to kill it by pumping in heavy mud.

On June 20, the drawworks disappeared into the muddy crater.

On June 24, three weeks after the well had blown out, a task force consisting of a full diesel drilling rig manned by Bakersfield crews from Gene Reid Drilling Inc. wheeled north from Kern County to Grizzly Island to try to kill the wild well.

The operation shaped up as a precise and potentially hazardous one. The crews' assignment was to drill a relief well to intercept and control the Kadane well that had gotten away.

The relief well at Grizzly Island would be, as far as anyone knew, the first one attempted in California, though such measures had been taken successfully in controlling Gulf Coast blowouts.

The relief well would be located 150 feet from the turbulent crater that marked the site of the Kadane well. With an estimated three million cubic feet per day of inflammable gas still bubbling up out of the crater, the new drilling operation would not have any margin for mistakes.

Changes in wind, which might carry the escaping gas to the relief well location, or fog, which would hold the gas close to the ground, would both be reasons for an immediate shutdown and possible evacuation from the relief well site.

The relief well was to be started in 20-inch conductor pipe which was being set at a depth of 50 feet. Crews were to drill straight hole to 100 feet, then whipstock in 12¼-inch hole to 500 feet, directing the hole toward the wild well.

After a depth of 500 feet was reached, the hole was to be opened to 17½-inch hole and heavy 13⅜-inch casing would be cemented solidly from that depth to the surface to serve as a strong anchor for blowout-prevention equipment. The blowout preventers that would then be installed would be those that might normally be used on a deep well—heavy double-hydraulic Shaffers and G.K. Hydril bag packer. After blowout prevention equipment had been tested to 1,500 pounds, the crews would drill 12¼-inch whipstocked hole to intercept the cavity at about 900 feet. An attempt would be made to kill the wild well with heavy 125-pound mud and quick-set cement with two percent calcium chloride in it to make it set up quickly.

The whole operation was to feature painstaking safety precautions. A Baroid mud wagon with a gas indicator to check mud for indications of gas would be used at the drill site, and an engineer was to be on 24-hour duty.

Five hundred barrels of heavy mud were to be on hand in special mix tanks. The rig itself—an Ideco 525 rated for 7,000 feet—besides being a full diesel featured all water-cooled exhausts and numerous other safety devices.

Men working on the relief well were to be covered by the contractor's normal insurance policy. As for the rig, the operator—G.E. Kadane & Sons—obtained the necessary insurance coverage.

On June 17, the Gene Reid crews, with Don Burge as toolpusher, spudded in to drill the relief well. Standing

Artist Paul Landacre said "Rig at Willows" depicted the elemental forces combating man's efforts to find and bring oil from the earth's interior. Less than a year later, the engraving could have served as a memorial to the fate of the Mighty Mole. (Brown Drilling Company)

guard between the men and the rig and the bubbling crater that marked the end of the trail for the Mighty Mole were a pair of gas sniffers—electronic devices that measured the gas content in the air and would sound an alarm if the content rose to a dangerous level.

On July 5, the gas blow was stopped by pumping in heavy mud and cement through the relief well. The relief well subsequently was drilled in straight hole to a depth of 6,621 feet and, failing to find anything at depth, plugged back for completion in the zone that had caused the blowout. The completion for a flow of 835,000 cubic feet per day of gas through a ⅜-inch bean from an interval at 654-665 feet was classified as a new pool discovery for the Suisun Bay gas field.

While Brown Drilling Company personnel were probing with a weighted line the muddy crater into which the Mighty Mole had disappeared and giving up the search for the lost equipment after they had played out 80 feet of line without hitting anything, another drilling contractor some 300 miles away was readying a new rig to take on an assignment that posed unknown pressures and hazards one might only guess at.

In Bakersfield, Clyde Hall of Clyde Hall Drilling Company proudly showed off the million-dollar rig he had assembled to drill a wildcat for Richfield Oil Corporation in the remote Yakataga area on the Gulf of Alaska. Hall took guests in an electric golf cart out to the corner of the yard where the completely winterized outfit had been rigged up.

A visitor said he had never heard of Yakataga, and Hall described it as "a tremendous metropolis of 12 people, give or take a few." He added, "You can buy a postage stamp but nothing else."

The rig that Toolpusher Vertis McDowell was taking north was a National 110 with 142-foot Lee C. Moore mast. It was rated for 20,000 feet with 4½-inch drill pipe. Its three Waukesha V-12 engines developed over 2,000 horsepower.

Even as the Mighty Mole was passing into history, the new rig began loading out for the long trip north to challenge the Alaskan prospect.

8 Looking for Opportunity

The aboveground nuclear tests that were conducted in Nevada in the 1950s, in spite of any effects they might have had otherwise, could hardly have been expected to have any bearing on the production of oil in California, but one producer claimed they increased his production.

The operator was Hugh Burns Hutchinson, who had come relatively late to the role of independent oil producer. Hutchinson, who lived in Long Beach, was a Scotsman who claimed Robert Burns as an ancestor and, in fact, had a seal he said had been used by the Scottish poet. He had begun his oil career working for Shell Oil Company in the Signal Hill boom in 1922 and later worked on construction jobs all over the world. including helping to build a refinery on Bahrain Island during World War II.

Like many other oilworkers, he had through a long period of years nurtured the dream of starting his own company. He had looked for the opportunity to go into business for himself and found it in 1954 in the form of a run-down lease in an all-but-defunct oil field at the foot of Conejo Grade on Highway 101 just outside Camarillo in Ventura County.

The 255-acre lease which Hutchinson took over was located some 325 air miles from the atomic test grounds in Nevada.

Hutchinson said the nuclear devices detonated in Nevada stirred up winds that were favorable to his operation. The reason the winds were favorable, Hutchinson liked to explain, was that he produced his wells with windmills.

There were six wells on the Hutchinson property at Conejo, ranging in depth from 30 to 100 feet. All were stripper wells producing from volcanics.

The name Hutchinson chose to run his one-man operation under was, after his middle name, Burns Oil Company. He abbreviated the name, after the fashion of such larger companies as Richfield Oil Corporation (Roco) and

Don Stussy, construction superintendent for Occidental Petroleum Corporation, with Lathrop Unit A-1, the discovery well for the Lathrop gas field. (William Rintoul)

Tidewater Oil Company (Toco), simply to Boco. The biggest insult he suffered relative to his modest operations, he said, came from his friend Dave Hardie, who was the manager for Bethlehem Supply Company's operations in the Los Angeles Basin. "All he ever says is 'Bunco,'" Hutchinson explained.

Production at Hutchinson's wells varied with the wind, ranging from as little as one-third barrel per day to as much as one barrel per day. Hutchinson produced the wells with all zones open. "Long before my time," he said, "some jarhead pulled all the casing out."

The reason for using windmills was a matter of economics. Almost as soon as he took over the lease, Hutchinson found that fuel costs were ruinous in lifting oil from the shallow wells. He noticed, too, that a brisk breeze blew up the slope almost every day.

A friend, W. Olsen, who was a sax player in the Harry Owens' band, gave him the first windmill. It worked so well he installed windmills at each well. The windmill wheels were 8 feet or 10 feet in diameter, except for one with an 18-foot wheel, which he purchased from a Camarillo orange grower.

He used conventional plunger-type pumps in his shallow wells. On the advice of Dave Hardie, he used ⅜-inch pipe for sucker rods, that is, for the pipe that formed the connection from the power source on the surface to actuate the pump in the well.

Windmills were not the only innovation. In one well, the fluid level was only seven feet from the surface. Hutchinson devised a "chain lift" to produce the well. The device consisted of an endless, wire, link-type chain that ran over a power pulley at the surface and around a pulley suspended some 30 feet down the hole.

He first used a 6-volt Dodge generator as a motor with a 6-volt Delco 150-watt gasoline generator supplying the power, but then hit on a more economical way. He put in a 70-to-1 reducer with an ancient one-lung gas engine that fired and then coasted for a period of time with no gas consumption. The gas engine turned the power pulley at 7 revolutions per minute, pulling the up-coming chain through a homemade stripper, that is, a device that peeled off heavy crude. At 7 revolutions per minute, Hutchinson found he could produce one barrel of fluid an hour, of which some 6 percent was crude oil and the remainder fresh, potable water.

One thing that puzzled Hutchinson about the lease was the varying gravities of the oil. From one well he might recover crude with a gravity of 13.5 degrees. From another of equal depth no more than 50 to 100 feet away, he recovered 17-gravity oil.

The crude cut 34 percent lube oil and 45 percent steam cylinder oil, making it suitable for use straight out of the well for tractor transmissions and truck differentials.

The oil that Hutchinson did not sell to farmers was trucked to the Edgington Refinery in Long Beach. Among other contracts, the refinery was a supplier of fuel for the Air Force. Hutchinson told friends he liked to think when a jet went roaring by overhead that perhaps due to some of his crude shipments, the flight would be smoother and speedier.

The chief beneficiary from the operation of his property, Hutchinson said, was the company for which he had once worked, Shell Oil Company. He drove from his home in Long Beach to the Conejo lease once a week, using his Shell credit card to pay for the gasoline he used for the approximately 200-mile roundtrip. The cost of the gasoline, he said, was the biggest item charged against operation of the lease.

Hutchinson's wells were the remnants of an estimated 110 wells that had been drilled in the Conejo field. Union Oil Company of California had discovered the field in 1892. Drillers had used almost every method known to man to put down the shallow wells, including spring poles, cable tools and rotary rigs. The average depth was only 150 feet.

When the depression hit in 1929, the Conejo field, a marginal operation at best, was virtually abandoned. Union gave up its lease, and in the years that followed various individuals leased rights to produce existing wells on what was known in the oil fields as a "poor boy" basis, that is, on shoestring capital with secondhand equipment, rusty tools and amazing ingenuity.

The field had produced 106,000 barrels of oil by the time Hutchinson put in an appearance and would produce about 4,000 barrels more before the last well finally was abandoned.

Hutchinson liked to entertain visitors with discussions of his plans for the Conejo property, ranging from multi-well drilling programs to putting in a roadside refinery where tourists could pump and refine their own oil for a fee.

His eyes twinkled when he spoke of one plan. "They put race horses out to pasture," he would say. "Why couldn't I let a lot of old pensioners, roughnecks and drillers come up

It was a "poor boy" operation, but for Hugh Burns Hutchinson it was fun producing his shallow stripper wells in the Conejo field. (William Rintoul)

to the lease and keep themselves interested doing odd jobs? For free, of course. I'd like to start with the old bunch of 1922 at Signal Hill with whom I worked—Crooks Stafford (Singer Jo Stafford's dad), Herbie Warren, Chet Inman, Eddie Fry, Charlie White and others too numerous to mention."

While Hugh Burns Hutchinson was finding opportunity in the harnessing of the wind to produce crude oil at Conejo, across the Coast Range mountains in Bakersfield events were moving to bring together two men who would transform an aging oil company few had heard of into one of the nation's biggest industrial corporations.

In the spring of 1957, Gene Reid Drilling Inc. received an invitation to bid on a shallow well to be drilled in the Compton Landing gas field four miles south of Princeton in the Sacramento Valley. The company that planned to drill the well was a firm named Occidental Petroleum Corporation, which had an office in Los Angeles.

Not having heard of the company before, Gene Reid made a few discreet telephone calls to supply and service companies to see if the company paid its bills. He learned that Occidental had been around for a number of years, had a small amount of production and did pay its bills. Reid bid the job, was the successful bidder and assigned the Hopper Slimhole rig he had put into service two years before to drill the well. The 2,600-foot hole on the Rheem-Zumwalt property proved to be a small producer, and Reid moved on to other drilling assignments for other operators.

It would be two years before Reid had any particular reason to think more of Occidental. In the meantime, there would be significant changes at both Gene Reid Drilling Inc. and at Occidental Petroleum Corporation.

On October 1, 1957, six months after drilling the Compton Landing well, Gene Reid addressed a letter headed "To Our Friends in the Oil Industry." In the letter Reid wrote:

"The present nature of general business conditions and in particular the depressed condition of the oil well drilling business in California has prompted the Directors of our

Corporation to give serious thought to its continued operation as an oil well drilling contractor.

"We enjoy a substantial net worth and our overall general financial condition is sound. However, it has been a matter of deep concern to us that contract income under present bidding does not provide the Corporation with a satisfactory return considering both variable and fixed costs.

"Gene Reid Drilling Inc., after careful consideration and pursuant to a plan adopted by its Board of Directors, has elected to liquidate its assets and dissolve."

In a personal note at the bottom of the letter, Gene Reid added: "I am one of the many who can be most appreciative and thankful for the wonderful opportunities and excellent treatment afforded me in the oil industry which has provided me with a measure of success. More especially, I am most grateful to my many friends in the industry and want to assure them that I intend to remain active as an individual operator in one form or another. I also hope to be able to contribute something of value to the California oil industry in the future."

The decision to get out of the drilling business ended a 15-year run for the contract drilling company, which had been formed in Bakersfield in 1942. The company had drilled an estimated 2,600 wells under contract to various operators, cutting more than eight million feet of hole.

On December 17, Gene Reid's five drilling rigs, valued at an estimated one and one-half million dollars, went on the auctioneer's block at 3505 Pierce Road. A spokesman for the auctioneer, Milton J. Wershow Company, described the sale as one of the largest oil well drilling spreads ever to go on the block in the United States.

The auction attracted 512 registered bidders and numerous spectators. Casey & Montgomery, Bakersfield drilling contractors, bid successfully for the five-acre yard and office that had been occupied by Gene Reid Drilling Inc.

The bids for the drilling rigs underscored the depressed circumstances of the drilling industry in California. The top bid for the contractor's Hopper Slimhole rig, which two years before had represented a $250,000 investment, was only $110,000. The bids for other rigs were similarly low, including an offer of $200,000 for a National 55 rig that had only been used for nine holes. Gene Reid rejected the bids and began negotiating the sale of the rigs.

Ten months later, the last of the five rigs that had been put on the auction block left Bakersfield, heading out to the Buena Vista field at Taft to handle an assignment for Honolulu Oil Corporation. The rig was an Ideco 525 with Bob Trout as toolpusher. It was going to work under Gene Reid Drilling Inc. After less than a year, Gene Reid was back in business with one rig, ready for action again.

In the meantime, there had been a significant change at Occidental Petroleum Corporation involving a graduate of Columbia University's School of Medicine who, among other things, had been a self-made millionaire in pharmaceuticals and cosmetics at 23, a representative in Lenin's Russia for the products of staunchly anticommunist Henry Ford's Ford Motor Company and 37 other leading American concerns, an art dealer and mastermind of the sale of the Romanov and Hearst art collections in department stores, the owner of one of the country's great breeding farms for Black Angus cattle and the proprietor of whiskey distilleries, including the J.W. Dant label that had been the biggest-selling Kentucky bonded bourbon in the United States.

The man was Dr. Armand Hammer, who had come to Los Angeles in 1956 to retire, having long since made his fortune. When his tax counsel advised him to find a money-losing tax shelter, he interested himself in Occidental, a 37-year-old company with a handful of employees and a market value of $120,000.

The acquisition of Dr. Hammer's interest in Occidental was negotiated in the lobby of the Beverly Hilton Hotel. Arthur Groman, senior partner in the law firm of Mitchell, Silberberg & Knupp in Los Angeles, one evening received a call from Dr. Hammer asking him to come to the hotel. In

the lobby, Groman found Dr. Hammer seated with a man whom Hammer introduced as J.K. Wadley. Wadley, wearing cowboy boots and a Stetson, was the owner of 11 wells in the Dominguez field 12 miles south of downtown Los Angeles. Occidental was interested in acquiring the wells.

During the evening, a deal was negotiated under which Dr. Hammer agreed to loan $500,000 to Occidental to finance the purchase of the Wadley wells in return for which Dr. Hammer was to get an option to turn his loan

Gene Reid was a wildcatter, always convinced that the next well would be the big one. (William Rintoul)

into stock at a fixed price of $1.50 per share. Groman borrowed a few sheets of paper from the desk clerk and wrote out the agreement between Dr. Hammer and Wadley in the lobby.

Soon afterward, Groman received another telephone call from Dr. Hammer asking him to come for a ride with him, explaining that Dr. Hammer wanted to "look at our oil wells in Dominguez." On the way, the two stopped at a drugstore because Dr. Hammer wanted to buy a Polaroid camera. When Groman asked what the camera was for, Dr. Hammer replied, "I've never seen an oil field before. I want to take some pictures."

In July, 1957, Dr. Hammer, who had become Occidental's major stockholder, was named president.

Having taken over, Dr. Hammer moved with characteristic vigor, but not, for awhile, with respect to Occidental. He acquired the ailing Mutual Broadcasting System for $700,000, tried for a year to get it back on track and finally sold it for $2 million.

Turning his attention to Occidental, Dr. Hammer concluded the company needed the expertise that could be furnished only by someone who knew the oil business inside and out, particularly the wildcatting end. He began making inquiries. The search led to Gene Reid in Bakersfield.

In July, 1959, Carl Blumay, who had just started working for Occidental as a public relations consultant, sat at his desk in an otherwise empty hallway which was the only space available at the company's suite of four small offices on Beverly Boulevard and wrote a press release which stated:

"Occidental Petroleum Corporation has merged with Gene Reid Drilling Inc., of Bakersfield (Calif.), it was announced jointly by E.C. (Gene) Reid, president of the firm bearing his name, and Dr. Armand Hammer, Occidental president.

"In the transaction, the two companies combined their various operations, organizations and assets, and Occidental Petroleum acquired the entire outstanding capital stock of the Bakersfield firm, all individually owned by Reid, in exchange for a block of capital stock."

The stock consisted of 160,000 shares, which made Gene Reid the second largest holder of Occidental shares, second only to Dr. Hammer. At the time, shares were selling for about $2 a share.

The release went on to state that Reid had been elected to the board of directors of Occidental and would serve as executive vice president in charge of all oil and gas production and development of the parent company.

"The combined organizations are developing a complete geolgical and exploration department to be supervised by E.F. (Bud) Reid, vice president of the Reid organization," the release concluded.

The merger involved less than a dozen employees, including Occidental's Dorothy Prell, secretary to Dr. Hammer; Paul Hebner, company secretary; Gladys Louden, accountant; and Gene Reid's staff, including his son Bud, the young geologists Dick Vaughan and Bob Critchlow, Reese Higdon, drilling and producton superintendent, Don Burge, toolpusher, Leo Adams, accountant, and Jean Peters, secretary. Vaughan and Critchlow had only recently joined the Gene Reid organization in anticipation of the merger. Soon afterward Geologist Bob Teitsworth and Engineer Charlie Horace joined the firm.

To the merger, Gene Reid brought the basic quality of the wildcatter: the conviction that the next wildcat would be the big one. "You've heard of the prospector who always figured the big vein was just over the next hill," Reid liked to say. "That's the way it is in this business. You've always got to see another foot of hole."

In the interests of seeing another foot of hole, Reid had spent an estimated $2½ million to drill wildcats in virtually every part of California, financing the exploratory work with income from his contract drilling business and from production he developed.

He had come early to the oil business, getting his first job in the oil fields at the age of 15 cleaning bricks and rebricking boilers on a lease near Maricopa in the Midway-Sunset

From the air it looked like a drilling boom with five rigs ready to spud in. It was far from a boom. In December, 1957, the oil industry in California was in a severe slump and Gene Reid Drilling Inc., for 15 years one of the state's leading drilling contractors, was selling its rigs. (Eugene F. Reid)

field. At the time, his father had been the Congregational minister in Maricopa.

To the nonwildcatter, the success Reid had enjoyed—a shallow find at Kern Bluff near Bakersfield, the development of leases in the Edison, Fruitvale and Midway-Sunset fields in Kern County—might long ago have provided the money for early retirement and a comfortable life of ease. To Reid, who had not bothered to move into a big new house or transfer his office to a swank executive suite, the money had meant a chance to drill more wildcats.

In looking ahead, however, he saw the wildcatter's role changing from that of the rugged individual who footed the bill alone to one of partnerships and joint ventures. He saw a new breed taking over, men like his son Bud, who held a master's degree in geology from Stanford University and had learned the ropes in the geological department of a major oil company, Shell Oil Company, before joining his father at Gene Reid Drilling Inc.

In the wake of the merger of Gene Reid Drilling Inc. and Occidental Petroleum Corporation, there was a clear and very simple division of responsibilities. Dr. Hammer's job was to supply the money. The Reids' job was to find the oil and gas.

The company chose to finance its exploratory activities through funding by investor/participants, who would put up the risk capital in return for shares of interest in whatever discoveries might be made. The company would supply the drilling rigs, the management and the skill, retaining a minimum of 50 percent interest.

A spirit of excitement permeated Occidental's operating headquarters in Bakersfield. "We could drill anywhere we wanted," Bud Reid said later. "It was a great feeling not to have to clear every decision with a committee or a board of directors. Instead of spending a lot of time fighting red tape, we spent it looking for oil and gas."

With a fresh outlook, the company decided to concentrate its efforts on the Sacramento Valley, a dry gas province overlooked by larger companies and relatively inactive for many years. Dick Vaughan liked the looks of the Arbuckle area 10 miles southeast of Williams, where Western Gulf Oil Company had made a discovery two years before.

Occidental negotiated a farmout with Gulf involving 34,000 acres, agreeing to drill a series of 10 wells on the fringes of the field. The number of wells was two less than the 12 already completed at Arbuckle. With characteristic enthusiasm, Dr. Hammer announced estimates of recoverable gas reserves under the farmout acreage at 700 billion cubic feet of gas, with an approximate value of $80

Dr. Armand Hammer was the financial wizard who gave up planned retirement to see what he could accomplish with little-known Occidental Petroleum Corporation. (Occidental Petroleum Corporation).

million after allowing for development costs and the possibility of 50 percent dry holes. The estimated reserves were approximately 10 times greater than the 68 billion cubic feet the state's Division of Oil & Gas estimated for the field.

Dr. Hammer gathered together a dozen or so friends to finance the drilling program, and Gene Reid's Ideco 525 loaded out to drill the first well. The hole was dry at 6,868 feet.

Occidental moved the rig to drill another hole and contracted with Brown Drilling Company to bring in another rig to step up the exploratory pace.

Eugene F. (Bud) Reid was the vice president and exploration manager for the Occidental Petroleum Corporation, which boldly set out to look for gas in the Sacramento Valley following the merger with Gene Reid Drilling Inc. (William Rintoul)

At the Arbuckle Unit No. W-1 on Sec. 34, 14N-2W, Colusa County, the Gene Reid rig found pay sand. A series of three tests of intervals from 5,523 to 5,852 feet yielded flows of gas totalling seven million cubic feet per day.

Dr. Hammer, with his pilot and dick Vaughan, flew up to the well in Hammer's Beech aircraft. On reconnoitering the short dirt strip at Williams, just outside Arbuckle, the pilot advised Dr. Hammer that he did not think he could take the plane down on that particular strip.

"Oh yes you can," Dr. Hammer said. "Go ahead."

Vaughan said later the pilot stopped only 15 feet in front of the rice check at the far end of the runway. Vaughan chose to take the Greyhound bus back to Bakersfield.

The Arbuckle find was the first of a series, involving the completion of extension tests at Arbuckle and new gas field discoveries at West Grimes in Colusa County in 1960 and at West Butte and Butte Creek in Sutter County in 1961.

The search led to a structure at Lathrop in San Joaquin County that was well known to geologists. Two major companies had drilled there, coming away without finding anything. Texaco had abandoned a 5,839-foot hole in 1937. Marathon had written off a 4,400-footer in 1947.

Bob Teitsworth decided the earlier wildcatters had not gone deep enough. Dick Vaughan and Bud Reid agreed. Noeth Gillette, Occidental's landman, leased 5,500 acres. The company moved in a Gene Reid rig.

The Lathrop Unit A No. 1 on Sec. 5, 1S-6E, San Joaquin County, located some 600 feet from the abandoned Texaco wildcat, on October 3, 1961, flowed 16,119,000 cubic feet per day of gas through a ¾-inch choke, getting the discovery in Upper Cretaceous E zone sands in the overall interval from 7,230 to 7,642 feet. The land block covered the entire field, which proved to be the second largest gas field discovered in California.

The discovery made possible in the following year the first payment of dividends to Occidental shareholders in 28 years.

It also enabled the company to explore for oil in the Los Angeles Basin, leading to discoveries in the Beverly Hills field and at Sawtelle, and provided an amount of capital sufficient to enable Dr. Hammer one day in 1964 to pose the question to his geologists, "If we could go any place in the world to look for oil, where would you want to go?"

The answer was Libya. For the first time, Occidental in 1966 would go overseas, embarking on a discovery trail that would take it beyond the deserts of Libya to the jungles of Peru and the stormy waters of the North Sea, finding more than four billion barrels of oil.

While Occidental Petroleum Corporation was getting ready to expand its search for opportunity to Libya, a new company was putting in an appearance in California in

Flare marked the successful testing of Occidental Petroleum Corporation's first well in the Arbuckle gas field. (Gene Reid)

1965 with a brighter prospect for the development of significant production than any company that had ever been formed in the state. The newcomer was as far down the production ladder as it was possible to get—it had no production at all—yet it was able to announce confidently, without challenge, that within three years it would be among the top ten oil producers in the United States.

THUMS Long Beach Company began life in the enviable position of having a property to develop where boundaries and reserves of crude oil—at least one billion barrels of it—had already been defined before the first well had been drilled. The property was the tidelands extension of the Wilmington field. It had been defined through formation studies of the onshore portion of the Wilmington field and of the Belmont Offshore field at the east end of the block, and through core drilling conducted by the City of Long Beach.

If the sailing was smooth, geologically speaking, for the THUMS play, the same could hardly be said for the seas that had to be sailed before the first well was possible. The first storm had come years before when the City of Long Beach, surrounded by oil production on three sides, refused to permit further encroachment of oil derricks within the city itself or within sight of its beach frontage. Development of new urban drilling techniques, including the soundproof rig and the landscaped hidden production "island," had breached this barrier.

Then there had been the matter of which branch of government—the City of Long Beach or the State of California—held title to tidelands oil. It had required a lengthy period of litigation and finally a complicated formula—providing shares to both parties—before this storm blew over.

And finally there had been the matter of subsiding land, which threatened to disappear beneath the sea with the continued production of oil from the onshore portion of the Wilmington field. This problem had been solved by finding a method that controlled subsidence—the injection of water at high pressure into oil-producing formations.

With barriers down, the city and state in February, 1965, opened the seaward extension of the Wilmington field, also known as the East Wilmington area, to competitive bidding, scheduling six days of sales for varying interests in a 4,600-acre tract estimated to contain 1,029,200,000 barrels of recoverable oil, believed to be the world's largest undeveloped proven reserve.

Preparations for bidding in some cases resembled sketches from spy stories and tales of intrigue. In the case of one joint venture, identical bids were sent from Los Angeles to Long Beach in four different cars following two

Occidental Petroleum Corporation's staff was small, but it was enthusiastic and willing to look for oil and gas where others had given up. Left to right, Bob Critchlow, Dick Vaughan, Charlie Horace, Bud Reid, Bob Teitsworth and Stan Eschner. Horace was the chief engineer, the others were geologists. (Occidental Petroleum Corporation)

A 14-mile gas line costing $2 million to build tied Occidental Petroleum Corporation's Lathrop gas field in with the Pacific Gas & Electric Company system, making Lathrop's gas available for the San Francisco Bay area and other PG&E users. This construction view showed a lineup crew "stabbing the pipe" and running stringer beads. (Occidental Petroleum Corporation)

different freeway routes. Each car contained a driver, security agent and company representative.

On the off chance that the car containing the required cashier's check might be delayed, arrangements were made to pick up another check in Long Beach for fast delivery to city hall.

In some cases, representatives carrying the bids were required to call headquarters just before bid-opening time for a last-minute check on the percentage to be offered. In one case, the bid figure was changed and a spare bid form, already signed, was filled out only minutes before bids were to be opened.

The first day's sale featured the big prize, an 80 percent interest in the tract. The successful bidder would automatically become the operator, designated as the field contractor, for the City of Long Beach in the development of the rich offshore reserve.

A crowd of over 200 jammed the Long Beach City Council chambers and an equal number watched proceedings over closed circuit TV in the Veterans Memorial Building auditorium one block away. There was a gasp of astonishment when Long Beach City Manager John R. Mansell read the first bid of a group consisting of Signal Oil & Gas Company, Standard Oil Company of California and Richfield Oil Corporation. The group offered 94.771 percent of the net profits for the right to develop the tract.

An unidentified man said: 'I thought it was going to be around 92 percent."

Another commented, "I guess they won it."

There was one other bid. It was even higher, a whopping 95.56 percent of net profits. The winning bid came from a five-company group which included Texaco Inc., Humble Oil & Refining Company, Union Oil Company of California, Mobil Oil Corporation and Shell Oil Company, known by the acronym formed from the first initial of each company's name as the THUMS group.

Both bids came as a pleasant surprise to city officials, who acknowledged that they had expected a high bid of about 92 percent. The state had been prepared to accept anything over 90 percent, Frank Hortig, executive officer of the State Lands Commission, said after the bid opening.

The remaining 20 percent working-interest shares were split into five offerings for the avowed purpose of guarding against monopoly and permitting small operators to bid. The offerings were each made on a separate day.

The five offerings included a 10 percent share for which the Pauley Petroleum Inc.-Allied Chemical Corporation group was high bidder with an offer of 98.277 percent of net profits; a 5 percent share for which the Standard Oil Company of California-Richfield Oil Corporation group was high bidder with an offer of 100 percent of net profits;

and offerings of 2½ percent, 1½ percent and 1 percent shares for which the Standard-Richfield group was high bidder with offers of 99.54 percent of net profits.

A study of winning bids showed that city and state governments would split 96.2528 percent of the profits from the seaward extension of the Wilmington field with oil companies sharing the remaining 3.7472 percent. Under the formula established for the division of those royalties belonging to city and state, the state would receive 85 percent, the city 15 percent.

Why had oil companies been willing to bid so high?

One reason lay with geography. The oil reserves were on the West Coast, a heavy importer of oil. They were near refining and transportation facilities. They were in the center of the largest petroleum-consuming region in the United States.

Following early successes, two grateful shareholders, left, surprised Occidental Petroleum Corporation officials with a gift of Oxy handkerchiefs, center, left to right, were Neil H. Jacoby, a director, Gene Reid, senior executive vice president, and Dr. Armand Hammer, president. (Occidental Petroleum Corporation)

Another reason was that the undertaking involved little financial risk. From profits, the field contractor would be reimbursed for investment and operating cost, and in addition receive a 3 percent management fee.

The THUMS group, which with the 80 percent share had won the right to be the operator, lost no time forming a new company to develop the seaward extension of the Wilmington field. THUMS Long Beach Company was staffed by loan employees from the parent companies, supplemented by a number of individuals employed directly.

The executives of the THUMS group chose Jack Russell to head the new firm. Russell, a 45-year-old former Navy pilot with wide experience in production operations, had

Rigs camouflaged to look like office buildings and extensive landscaping with palms and other semitropical trees made the seaward extension of the Wilmington field an attractive neighbor. (THUMS Long Beach Company)

for the preceding four years been manager of producing operations for Mobil's Los Angeles exploration and production division. He was a graduate of the University of Texas with a bachelor's degree in petroleum engineering and had been with Mobil 20 years.

On July 16, a group of oil company executives, state officials and Long Beach city officials gathered on Pier J to watch G.A. Burton, vice president of Shell Oil Company and chairman of the THUMS board, push a button to signal the start of drilling the first well.

As a horn sounded, the Camay Drilling Company crew lowered the bit and started the kelly turning to the right at No. J-145, which was the first of a scheduled 1,100 new wells. The well subsequently was completed for 1,104 barrels a day net from total depth of 4,562 feet.

Before the year ended, THUMS had begun construction of two of four man-made islands from which the major share of the tidelands tract would be developed. The islands were located in 30 to 40 feet of water, with the surface rising 15 feet above water. The first was nine acres in size, the second ten acres.

To build the islands, THUMS needed a big dredge. The company located one in Canada and brought it to California by freight train. It took 170 freight cars to accomplish the move. The dredge pumping machinery was believed to be the most powerful in the world, driven by a 10,000-horsepower electric motor. The dredge sucked up sand from the bottom of the harbor in areas approved by the Army Corps of Engineers to form a base for the islands. Barges hauled rock for the perimeter from a quarry on Catalina Island some 22 miles away, and the dredge filled the inside.

Drilling began from Island Alpha in 1966 and from Islands Bravo, Charlie and Delta in 1967, on schedule. By the end of August of that year, 19 rigs were drilling from the four islands and from Pier J, making East Wilmington the busiest field in the United States.

The contractors handling the drilling work included Moran Brothers, Wichita Falls, Texas, four new all-electric

rigs on Island Alpha; Island Drilling Company, a combination of Commonwealth Drilling Company of Canada, Signal Drilling Company, Denver, and Brinkerhoff Drilling Company, Denver, four new electric rigs on Island Bravo; Camay Drilling Company, Los Angeles, four new electric rigs on Island Charlie, two on Island Delta and one on Pier J; and Big Chief Drilling Company, Oklahoma City, four rigs on Island Delta.

By the end of the year, THUMS was completing wells at a rate of 350 per year, or almost one a day. Production had climbed past 50,000 barrels a day and was continuing to climb.

That same year, the Long Beach City Council approved a new name for the group of man-made islands, calling them the Astronaut Islands. Individually, they were to be known as Grissom, White, Chaffee and Freeman in memory of the astronauts who lost their lives.

Night lighting of the Astronaut Islands led some to describe the THUMS Long Beach Company development as the world's most beautiful oil field. (THUMS Long Beach Company)

To ensure that the offshore oil operation did not offend anyone, THUMS engaged the architectural firm of Linesch & Reynolds to suggest a beautification scheme that would camouflage and soundproof the operation. The resulting program included constructing the drilling towers with pastel balconies to look like office buildings or penthouse apartments. Mounted on rails, the towers ranged around the islands to drill slant wells six feet apart. Each tower had a triple deck substructure to house mud pumps, shakers for drill cuttings, power converters, motors and other equipment.

To mask the view of drilling equipment, THUMS planted the islands with semitropical trees: Washingtonia fan palms and Canary Island date palms. For lower plantings, climate-resistant sandalwood trees, salt bushes, oleanders, acacias and Moreton Bay fig trees were used. The ocean-dredged sand had to be washed of salt and beefed up with wood shavings and fertilizer. An elaborate water and fertilizing system was installed.

As the first island began to shape up, Long Beach citizens became enthusiastic. Back to the drawing boards went planners to come up with three spectacular 30-foot-high waterfalls, visible from all around the bay. It looked so good the city fathers demanded, and got, floodlights for night viewing.

9 The Asphalto Play

The idea there might be an undiscovered oil field at Asphalto struck Hy Seiden, a consulting geologist, with an impact he would later describe to friends as having caused him almost to jump out of his skin. It was a hot afternoon in June, 1961, and Seiden, a compact 5-foot 5-inch figure in polo shirt and slacks, was sitting in the office he had fashioned, competently but not with professional skill, from what had been the garage of the tract house he was buying in Bakersfield. He had been poring over the electric logs of three wildcat wells in the same manner a playwright might mentally kick around the facets of a broad theme, hoping for the germ of a smash hit to spring to mind.

The electric logs were graphlike charts, readily available through oil industry sources to anyone interested. Made at the time the wells were drilled, they recorded in wavy lines the reaction of formations through which the wells passed to electrical impulses generated through a device run into the holes on conductor cable. To the trained eye, the logs described the types of rocks the wells had penetrated. Other data revealed the direction and to what degree the formations dipped, vital information to the oil-seeker.

The wells had been drilled by different oil companies. They included E.A. Parkford's Parkford-McKittrick No. 1 on Sec. 22, 30S-22E, Kern County, which had gone to 7,732 feet in 1947; Sunray Oil Corporation's Leutholtz No. 1 on Sec. 22, 30S-22E, Kern County, which had gone to 6,547 feet in 1948; and Standard Oil Company of California's Van Wert No. 68 on Sec. 26, 30S-22E, Kern County, which had gone to 6,098 feet in 1951. Each company had looked for an oil field at Asphalto, 30 miles southwest of Bakersfield. None had found one.

The first optimistic flush passed. How could an undiscovered oil field possibly exist in an area picked over as thoroughly as Asphalto? The abandoned mining camp that gave the treeless, sagebrush-covered area its name had come into being in the 1860s when transplanted Mother

The Asphalto discovery came at a time when California's wildcatters badly needed a boost. It proved there was more oil to be found, and before the dust had settled in the McKittrick area, more than 100 million barrels of additional oil had been discovered. (William Rintoul)

Lode mining men dug asphalt from deep shafts, refined it over open fires fueled in part by the oil-impregnated bones of sabre-toothed tigers and other prehistoric animals found with the asphalt, and dispatched the finished product in wooden crates via 24-mule teams to a railhead in Bakersfield, there to be shipped north to pave the streets and sidewalks of San Francisco and grease logging skids in Northern California timberlands.

Since those days, wildcatters had drilled the area for miles around the crumbling shafts and eroding mine dumps of Asphalto, finding three major oil fields. To the east, hardly more than a mile from where Seiden's brainstorm told him new oil might be found, lay the Elk Hills field, one of only a dozen fields in the United States with enough oil to eventually produce one billion barrels. To the

Ruth Seiden with Richard, left, and Stuart at the well that proved up the Asphalto field. (Hy Seiden)

south, less than two miles away, lay the sprawling Midway-Sunset field, another of the billion-barrel fields. To the west, two miles distant, lay the McKittrick field, which, though not as large as Elk Hills or Midway-Sunset, still ranked as one of California's major oil fields. Could everyone have overlooked another oil field at Asphalto?

"Ruth," Seiden called.

A moment or two later, his wife appeared, wiping her hands on her apron. She looked as if she knew what to expect and was anxious to get it over with so she might return to the kitchen.

"Look at this," Seiden said, with the air of a man who has just found a gold nugget on a busy California street corner. He pushed the logs aside and spread before her a map of the Asphalto area, showing the oil fields that had been found and the abandoned wildcats that had been dry. "See these dry holes?" He pointed at three symbols on the map: small circles with crosses drawn through them. The three circles formed the points of a rough triangle, one-half mile long on one side, two miles long on the other sides. "Some-where in here," Seiden said, stabbing his finger into the center of the imaginary triangle, "there may be an oil field."

Ruth, the former Ruth Goldstein of Yonkers, New York, a onetime stenographer who in 13 years of marriage had absorbed a layman's knowledge of geology, listened while her husband described the clues.

The dipmeter for one well indicated the formation known to geologists as the Antelope shale, which often contained the Stevens sand, a prolific source for oil in the nearby Elk Hills field, dipped to the south-southwest. In another well nearby, the formation dipped to the south, suggesting the possibility of a structural nose—a geologic feature that often forms a trap for oil. In the third well, the

Though he had put up money for nearly one hundred wildcats and never hit it big, E.A. Bender had a special regard for the area where Hy Seiden wanted him to drill. Above, Bender with the big one that did not get away. (William Rintoul)

zone dipped to the north, suggesting the far side of a channel in which sand could have been deposited. In each well, the formation was a shale body with thin layers of oil-stained sand. These layers were to Seiden potent clues.

Challenged by his wife's lack of understanding, he tried harder to explain. He mainly succeeded in getting himself more excited. The shale, he said, could contain the Stevens sand, which would have been laid down in a channel some 12 million years ago. It looked as if the channel ran from northeast to southwest. If the channel were penetrated by a well at a point where the sand crossed the contours of the structural nose to form a trap, the sand might give up oil.

By the time Ruth had excused herself to return to the kitchen and, in leaving, congratulated her husband for his idea, adding the suggestion he should perhaps work it out further, Seiden had become so sold he knew he would have to forego working on anything else. There would be much to do. He would have to get all the information he could find on the geology of the Asphalto area, interpret it, decide where a well should be drilled and get the land to drill on. In short, he had to put together what in oil field phraseology is known as a "play," the creative groundwork that precedes the drilling of an exploratory well. Afterward, of course, would come another hurdle. He would have to sell the play.

Hy Seiden, 46 years old, had taken the risky plunge into consulting six months before, after nine years as a geologist with Standard Oil Company of California. His interest in geology went back only a few years farther to Army days during World War II.

Born in New York City, the sixth of eleven children of Harry and Celia Seiden, Seiden grew up in Yonkers, New York, a manufacturing center far removed from oil fields. His father, a painting contractor, and his mother, a dressmaker, had helped finance their children's educations through bank loans. When Hy's turn for college came in 1933, the banks were closed. Hy had a paper route and he continued working it for two years before entering Yonkers College, where he spent two years in pre-med.

His money ran out and Seiden worked at odd jobs, trying to save enough to go back to college.

As the United States was drawn toward World War II, Seiden enlisted in the Army, arriving as a coast artillery-man at Fort Kamehameha on the Hawaiian Island of Oahu six months before the Japanese attack on Pearl Harbor. Later he attended Officer's Candidate School at Camp Davis, North Carolina, earning a commission in anti-aircraft. He put in three years in Panama, leaving the service a Captain.

On leave in the summer of 1944, Seiden ran into a friend from Yonkers who was working in Illinois as a geologist for the United States Geological Survey. The friend explained what geology was about, how it was a study of the earth, its rocks and its origin. To Seiden, geology before had been little more than a word; the earth's rocks had been, he recounted later, "something you threw and broke windows with." He found the friend's explanations so interesting that when he returned to college early in 1949—to the University of California at Los Angeles—he sandwiched in an introductory course to geology.

Among other things, the course touched on economic aspects of geology. It occurred to Seiden geology could be of more than academic interest. He investigated job oppor-tunities and found the largest number of geologists, some 90 percent, were employed in the petroleum industry.

In the summer of 1950 he graduated from UCLA with honors and a bachelor's degree in geology. A year's gradu-ate study preceded a friendly interview with a recruiter for Standard Oil Company of California. The interview led to a job offer, and Seiden soon afterward found himself working as a geologist in the San Joaquin Valley, a huge sedimentary basin containing some of California's most fertile farm land and many of its largest oil fields.

At the oil company, Seiden broke in working on subsur-face geology and doing field mapping. He divided his time between air-conditioned offices in Coalinga and Bakers-field and the field, where he spent long days mapping geologic features. The men with whom he worked were much like himself, graduate geologists with an abundant interest in the work they did. They were members of a team, working to find oil.

The first years brought regular advancement in both responsibility and salary. After awhile, however, advancement became more difficult, work projects seemed stereotyped. A vague dissatisfaction began to gnaw at Seiden. On one hand, he thought, was the chance to be a member of a team that tackled big projects, the pleasant association with other geologists, the security of a regular pay check. On the other hand, what of the ideas he had that he could not develop, the plays he would like to work up that did not fit the company's plans, the areas he would like to investigate that the company did not favor? After long soul-searching, he reached a decision. In January, 1961, with neither nest egg nor long list of clients, with two sons to support, Stuart, seven, and Richard, three, he opened a consulting office at home.

In six months of consulting, Seiden picked up no steady retainers. He worked up a small play in the McKittrick area and managed to sell it, getting a fee of $2,000 and 10 percent interest in any oil the well found. The well was dry; the money soon gone. Ruth talked of going to work. It might finally be necessary for Ruth to get a job, Seiden said, but first he would do what he could with part-time jobs. He said he would like to give himself a few years and some good oil plays before giving up the idea of working for himself.

It took almost a year to put together the Asphalto play. Normally it would not have taken so long, but because of odd jobs, Seiden was only able to work on the play part-time. While at UCLA, he had passed most requirements for a teaching credential. This qualified him for substitute teaching, and he pinch-hit in Bakersfield's junior highs and high schools, teaching English and physical science, his minors, and other subjects when required. He earned $20 a day substituting. While the money was not as good as at Standard Oil Company of California, he tried hard to keep from wishing more teachers would get sick.

He picked up spare change selling potato chips, pretzels and salted nuts for a Los Angeles distributor. When that proved less than profitable, he tried his hand selling metallic covers for home swimming pools. Another part-time job was the booking of stage shows. His sister, Ginny, sang professionally under the name Ginny Barry; his brother-in-law, Dave Barry, was a comedian whose credits included

Storage tanks sprouted overnight at Asphalto within view of the diggings where Mother Lode gold miners had sought asphalt one hundred years before. (William Rintoul)

appearances on the Ed Sullivan show and in such movies as *Some Like It Hot,* where he played the manager of the all-girl orchestra featuring Marilyn Monroe. Working with an agent in Los Angeles, Seiden booked shows for the Greater Bakersfield Trade Club, the Wing-Ding parties of the graduating classes of Bakersfield high schools and for various benefits.

Every free minute went into working up the Asphalto play. Seiden rounded up all the information he could get—from the state's Division of Oil & Gas publications, geological bulletins and from other geologists and companies. He used the information to build a picture of the earth beneath the surface: the layers of formations, the structure, the distribution of the sand that might contain oil. It was a detailed research problem.

After months of study, Seiden deduced the target—the Stevens zone—would require a depth of perhaps 6,000 feet for evaluation. To drill that deep would cost perhaps $60,000—a large sum for a look at something that might not even be present, much less productive of oil. There was only one way to find out. The next step was to get the land. Seiden looked at a title map. Every acre was already spoken for.

The first land owner Seiden approached was Lester C. Hotchkiss, a veteran broker who lived in Fresno and leased government land, then turned the leases to responsible operators to secure the drilling of exploratory wells. The lease broker, a sandy-haired man of medium-height, gave Seiden a friendly reception, going so far as to buy him lunch. If Seiden could get someone to drill, Hotchkiss said, he would be happy to make a deal, putting 240 acres into the land block. He and Seiden would divide the lessor's interest and split whatever bonus Seiden might receive if and when he sold the play. Seiden explained his situation with regard to ready cash. The broker offered to take more of the bird in the bush—the lessor's interest in the acreage—and let Seiden keep all of the bird in the hand—the hoped-for bonus when the play was sold. The agreement was verbal.

The discovery of the Asphalto field revived interest in the whole McKittrick area, leading to a series of discoveries in such areas as the Northeast area of the McKittrick field, where Union Oil Company of California completed a 6,389 barrels-per-day well. Admiring the company's Lowell-Wible No. 47X were, left to right, Dick Glass, drilling foreman; W.P. (Bud) Knick, area superintendent; Henry Barnes, field operator; Ed Borglin, exploitation geologist; and Bill Spradlin, production foreman. (William Rintoul)

The next land owner Seiden approached was Intex Oil Company, a small independent oil company with offices in Bakersfield. Russell Gough, the company's land agent, agreed to go into a deal as soon as Seiden lined up someone to drill. The company would put up 240 acres, giving half interest in its lease in return for the drilling of a well on the company's land with no expense to the company. For the company, it was a good chance to find what lay beneath the surface without spending money to drill. Again, the agreement was verbal.

One more hurdle remained. Standard Oil Company of California, the major oil company for which Seiden formerly worked, owned acreage that appeared to be in the heart of the projected oil field. Without the company's cooperation, it would be impossible to interest anyone in drilling. If no one drilled, the rest of the flimsy stack of cards—the land promised from the lease broker and the independent—would come tumbling down.

On a fall afternoon Seiden sat in the office of Standard's land manager, Bill Croghan, asking for the chance to drill on the company's land. The deal would be advantageous to all concerned, Seiden said. The company, with no expense to itself, would find out what lay below the surface. The land manager listened attentively. His company had drilled in the area once without finding oil. It was doubtful if the company would soon drill again. True, the company would have to give up a portion of a bird in the bush—a percentage of whatever oil might lie under the small portion of land the company farmed out—but the company also stood to gain if the well hit, not only through the 50 percent the company would keep but also by drilling on land the company retained. And even if the wildcat did not find oil, it might develop geological information that would lead to an oil discovery at a more favorable site. The company agreed to put up 80 acres, giving Seiden a block that totalled 560 acres.

The block was none too big. Seiden had misgivings about selling the play. But there was no turning back. Someone had to be found who would risk $60,000 to drill a well, gambling the money on Seiden's judgment. It would be a gamble in which the odds of finding a commercial amount of oil, according to statistics of exploratory drilling, were approximately 3 in 100.

The offices of men who had $60,000, or access to it, were in some respects similar. Most contained framed pictures of wooden derricks harking back to early days of oil, stacks of technical publications and rolled maps pigeonholed near drafting tables. These were the working offices of men who look for oil.

Hy Seiden brought with him the folder that contained the Asphalto play. The language of the report was neither flowery nor stilted. It was, like the language of much classic literature, simple and to the point, describing the oil field Seiden hoped to find, the clues that indicated the field to be present and the economics of drilling for the field. Along with the report were maps showing as precisely as Seiden could visualize what a well drilled at Asphalto might expect to find: the thickness of the formations through which the well would pass, the depth at which oil might be found.

If the offices bore similarity, objections to the play might be described as similar too: not enough indication the Stevens sand would be present, let alone productive of oil; dry holes too close to where Seiden proposed to drill; land block too small. The play proved hard to sell. Seiden continued with his part-time jobs—he was selling pool covers at that particular time—while he tried to interest somebody in drilling the Asphalto play.

Among other things, the climate was discouraging for exploratory drilling in California. There had been a dearth of onshore oil discoveries since the days of the Cuyama Valley finds more than 10 years before. In the wake of Cuyama, wildcatters had swarmed into Carrizo Plains, an empty forbidding area that lay between Cuyama's booming fields and the prolific fields of the southern San Joaquin Valley. There they had broken their picks, drilling an unbroken string of dusters. Between Carrizo and the lack of success elsewhere, shadows seemed to be lengthening over the exploratory scene in California.

One of the men to whom Seiden showed the play was a Bakersfield wildcatter named E.A. Bender. He had begun his oil career almost 40 years before, working in a pipeline gang at Elk Hills, near where Seiden said there was a chance to find oil. Bender, a tall, energetic man whose inventive mind had come up with the inspiration for many new drilling tools, through the years had put up money for close to one hundred wildcats, most of them dry, without hitting it big.

He liked the play, but for more reasons than geology. Years before, while Bender had been working in the pipeline gang near McKittrick, he had been called aside one day by the foreman, who by his manner made it plain he had something important to discuss. Bender had been fearful

Three weeks after Union completed its blockbusting Lowell-Wible No. 47X, Standard Oil Company of California brought in a well one mile to the west, flowing at a rate of 9,000 barrels per day. Claiming championship honors for the company's 7Z No. 558 were, left to right, E.H. Ransom, drilling supervisor; G.E. Furman, production foreman; R.A. Geissel, drilling foreman; Fred Kalenborn, district superintendent; and Charlie Moncrief, division manager. (William Rintoul)

the foreman meant to fire him. Instead, the foreman complimented him on his willingness to work and gave him a promotion. Bender had decided the sun-baked area around McKittrick was a lucky one for him.

Though Bender liked the play, he felt the price tag and the depth were steep. He began a search for friends who might share the financial risk in return for a share of the oil, if any. Prospective backers shied off. Bender faced the age-old dilemma of the wildcatter. Should he back off or go ahead?

On November 17, 1962, E.A. Bender spudded in with a contract drilling rig from Hagestad Drilling Company to drill the well that Hy Seiden said might tap an oil field. The well was designated as Standard Oil Company No. 18 on Sec. 23, 30S-22E, Kern County. For 13 days, Seiden hovered at the well site, watching mud samples coming from the hole, living in a state he later described as "calm nervousness."

On the 13th day, the drilling bit penetrated Stevens sand. Mud coming from the hole bore oil shows. A logging device was run. The zone looked promising. Another device was run to take samples of the formation. The samples looked good. A testing truck was called out. The test came late at night.

"I stood there and waited," Seiden recalled. "After three or four minutes I smelled gas coming out." A hose ran from the test pipe to a bucket of water, bleeding off a sampling. The bubbling in the bucket indicated something was coming into the tester. "It began to bubble strongly, indicating a good blow. After about nine minutes oil started flowing. I was dazed." Someone filled a bucket with oil. "Man, it smelled good."

On December 14, 1962, the discovery well was completed flowing 312 barrels a day of 36-gravity oil and 825,000 cubic feet per day of gas through a 12/64-inch bean. The productive interval in the Stevens sand was from 5,612 to 5,762 feet.

Overnight the Asphalto oil field ceased to be the brainstorm of a persevering geologist and became instead the brightest oil discovery of recent years in California.

The well that E.A. Bender drilled on the basis of Hy Seiden's play would prove to be the first of more than 80 wells that eventually would be completed in the field. The oil the well produced would prove to be the first installment on some 35 million barrels the field would eventually produce.

The discovery of the Asphalto field revived interest in exploration in the McKittrick area, giving wildcatters new targets at which to shoot, and new concepts with which to pursue additional discoveries. In the wake of Asphalto, there was a string of new discoveries, including the discovery of the Railroad Gap field in 1964 by Standard Oil Company of California, deeper pool discoveries by the same company in the McKittrick Front and Cymric Flank areas of the Cymric field in 1965-1967 and deeper pool discoveries by Standard in the Northeast area of the McKittrick field in 1964-1965 and by Rothschild Oil Company in the Main area of the McKittrick field in 1964. The discoveries proved up more than 100 million barrels of oil.

The rapid expansion of the Asphalto field after the discovery brought production to seven companies, including major companies and independents. Among the latter was Bob Ferguson, a 35-year-old geologist from Long Beach.

In June, 1962, five months before E.A. Bender moved in to drill, Ferguson had leased 320 acres of federal land. Since the land at the time was not on any known geological structure, Ferguson had leased it for a total of $170, including a fee of 50¢ an acre, or $160 for 320 acres, and a $10 filing fee, with 12½-percent royalty retained by the government.

At least two major oil companies had held the property. Standard Oil Company of California had drilled the Van Wert duster there. Union Oil Company of California had picked up the lease but had given it up rather than risk a well. Ferguson was one of some 90 who filed for the lease when it came up in the government's lottery; his name had been drawn.

After the Asphalto discovery, operators beat a path to Ferguson's door seeking farmouts. Ferguson subsequently assigned 40 acres to J. Ainslie Bell, Operator, who completed four wells. Ferguson turned 80 acres to Sunray. The company brought in eight wells.

With hefty royalty payments from Bell and Sunray, Ferguson entered the operational ranks on his own, proceeding to complete 10 wells. Less than one and one-half years after he had risked $170, Ferguson had qualified for admission to the ranks of one of America's most exclusive clubs—those making over $1 million a year. According to the Internal Revenue Service, the "club" listed something like 398 members.

For Hy Seiden, the Asphalto discovery, of course, meant a crowning success. One afternoon about a year and a half after the discovery, he sat in the same office where he had conceived the idea Asphalto might exist, surveying a scene not much changed from that he saw the afternoon the idea hit.

The office was perhaps a little more cluttered. As royalties from Asphalto began to come in, he indulged an appetite that had been held in check for years. An avid reader, he bought many books. The titles on the shelves ran a wide variety, ranging from a complete set of Shakespeare—one of his first purchases—to the latest selections of the Book-of-the-Month Club.

On his desk, the same desk, sat a shiny clock-and-pen set, featuring a golden derrick, a clock and a pen. The San Joaquin Geological Society awarded the set to Seiden for presenting the best talk given during the Society's 1962-1963 meeting schedule. The title, naturally, was "The Asphalto Field."

In looking back, Seiden made no secret of his gratitude. "Asphalto enabled me to stay in geology and oil exploration," he said, "which is the thing I prefer to do. It's a fascinating business."

Rig at twilight at Asphalto. (William Rintoul)

10 A Hotfoot for Lazy Oil

At 1:30 P.M. on a Thursday afternoon in July, 1956, John H. Thacher Jr., vice president of California Research Corporation, a subsidiary of Standard Oil Company of California, turned a knob on a small boxlike device to set on fire an oil sand some 2,000 feet below the surface where he knelt. The scene was Sec. 5, 11N-23W, Kern County, two miles east of Maricopa in the Midway-Sunset field.

The control which Thacher had switched on activated a heating coil opposite oil sand at the bottom of an adjoining well. Air earlier had been injected into the sand to build up a mixture of gases suitable for combustion.

Three minutes later, the announcement was made that the temperature rise transmitted to the surface by a thermistor—a small, rugged temperature-measuring device at the bottom of the well—confirmed that the underground fire had been successfully started. Later, it would be rumored that charts indicated spontaneous combustion actually had occurred hours before, caused by the heat generated by the oxidation of oil in the reservoir in response to the injection of air.

While those on the surface perspired in 96-degree summer heat, two thousand feet below them the temperature climbed to the more than 1,000 degrees Fahrenheit which would accompany burning of the oil sand.

Some of those on hand for the start-up of the project sought relief from the sun in a doghouse that doubled as an office and change room. A potted plant with florist's foil wrapped around the container added a homey touch. It was a flame bush, a rare bush which grows in Southern California, producing a fiery bloom similar to a flame. It had been presented to Pete Simm, senior research engineer in charge of the project, by his wife and children with a verse paraphrased from a song titled "I Don't Want To Set The

Ernie Gollehon, Halliburton thermal operator, on a job for Standard Oil Company of California on the Monte Cristo lease at Kern River, June 1964. (William Rintoul)

World On Fire," which had been popular several years before. The verse read:

"We don't want you
   to set the world on fire.
We just want to keep
   the flame in your heart."

A few weeks earlier in the South Belridge field 30 miles northwest of where John Thacher had turned on the heat, General Petroleum Corporation had ignited an oil sand. The company's engineers had designed an elaborate tool to ignite the sand by electrical impulse. The tool was so impressive they had patented it. To set the stage for use of the tool, they had begun injecting air to build up a mixture that would support combustion. Oxidation of oil in the reservoir generated heat, and spontaneous ignition had occurred before the tool could be used.

The two projects—General Petroleum's at South Belridge, California Research's at Midway-Sunset—were the first firefloods in California, and the most ambitious that

On a hot July afternoon, John H. Thacher Jr. activated a heating coil to set on fire an oil sand some 2,000 feet below the surface. It was part of an attempt to increase recovery of heavy oil by literally boiling the oil out of the ground. (William Rintoul)

had ever been undertaken anywhere. They followed small pilot tests conducted by Magnolia Petroleum Company, a General Petroleum affiliate, in the West Loco field in Jefferson County, Oklahoma, and by The California Company, a California Research affiliate, in the Irvine-Furnace field in Estill County, Kentucky.

It was all part of a continuing and determined industry-wide effort to solve a perplexing problem, or at least ameliorate it, that was well known to oil producers but little understood beyond oil's ranks. To a public whose knowledge of oil stemmed largely from Hollywood movies, the whole challenge was to find the oil, with no thought that there might be any problem afterward in producing it. There seemed little more to the oil business than taking the risk of drilling a well, which was either a gusher with oil blowing over the top of the derrick or a dry hole. If the well was productive, it was assumed that oil would flow until the well ran dry, or perhaps until it was necessary to put the well on a pump. In either event, it was taken for granted that the well would be produced until there was no more oil left, in other words, until all the oil that had been found had been taken from the ground.

For oil producers, there unfortunately was more to it than that. The truth was, they lacked the tools and techniques to recover the major share of the oil they found. It was a harsh fact of life that for every barrel they were able to bring to the surface to convert into the products that fueled America's economy, they would leave at least two barrels in the ground.

The situation was even worse when it came to much of the oil that had been found in California. The oil was what was known as heavy crude, by general definition, oil of 20 degrees gravity or less, based on a scale developed by the American Petroleum Institute to express the density of liquid hydrocarbons. In the scale, water had a gravity of 10 degrees.

Depending on whose figures one took, wildcatters had found as much as, if not more than, 40 billion barrels of heavy crude in California. With conventional techniques, it appeared that they would be able to recover only a small percentage, probably no more than 10 percent.

The crude was thick and sticky and moved like molasses on a cold day. In the Kern River field at Bakersfield, one of the state's biggest heavy crude fields, the 13-gravity oil came up from the earth looking as black as tar, and scarcely more mobile. One could poke a stick into the sluggish crude, then hold the stick away and watch a thin ribbon of the black oil fall slowly to the ground. In the Santa Maria Valley field at Santa Maria, some wells tapped sand whose oil was so heavy, ranging from 6 to 9 degrees gravity, that the crude could only be produced by mixing it with diluent, a 21-gravity oil piped in from another field. The diluent was pumped down the wells to mingle with native crude, lightening the crude so that it might be pumped, together with the diluent, to the surface. The crude in its native state appeared so solid that one might reasonably have expected it to be mined, rather than produced through a well.

Though heavy crude moved slowly, if at all, it could be made to flow under the proper circumstances almost as smoothly as water. The circumstances involved the application of heat. An apt analogy was with what happened when one took butter out of the refrigerator and put it in a frying pan. What in the first instance might have seemed to be a solid would, with the application of heat, become thoroughly fluid.

The setting on fire of sands in place at South Belridge and Midway-Sunset represented one approach, a direct one, to the matter of applying heat to heavy oil to make it flow more readily to producing wells through which the oil could be recovered.

Essentially, both experiments consisted of starting and perpetuating a controlled fire in subsurface oil sand to reduce the viscosity, thus increasing the mobility, of the oil with heat and drive the oil ahead of the burning front toward producing wells. To support combustion, it was necessary to continue injecting air into the oil sand.

Keeping close tabs on the control manifold which regulated the flow of compressed air and gas to the injection well at the California Research Corporation fireflood in the Midway-Sunset field, were, left to right, Joe Lindsay and Bob Dunlap, research assistants. (William Rintoul)

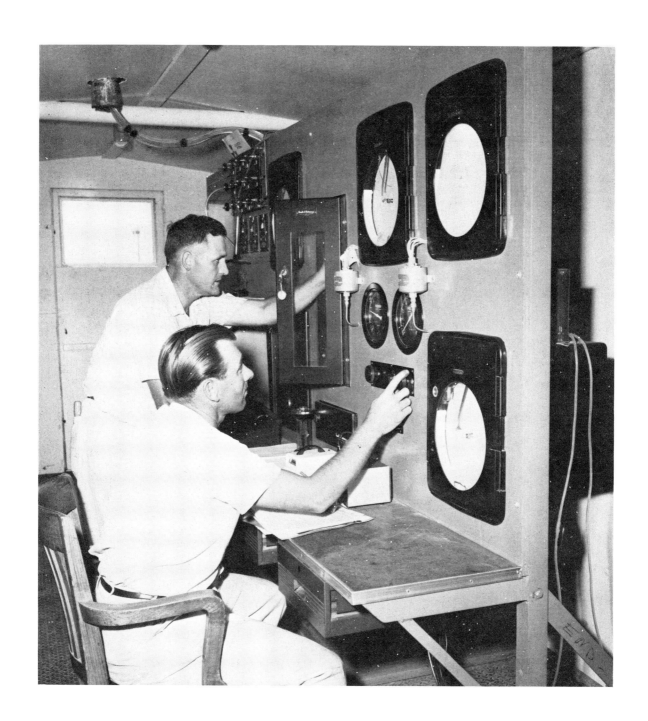

A Hotfoot for Lazy Oil   209

From a conservation standpoint, it was anticipated—correctly, it developed—that no more than 10 percent of the oil would be burned, and that would be mostly coke and heavy residuum, the least usable part. The lighter, usable oil would be pushed ahead of the burning front to producing wells.

Field work for the General Petroleum project had begun in the summer of 1955. The site selected for the experiment was the Marina lease on Sec. 10, 29S-21E, Kern County. The sand was approximately 30 feet thick and occurred in the upper portion of the Tulare zone at about 700 feet. The company utilized a five-well pattern consisting of four producing wells 330 feet apart and an injection well in the center.

Engineers anticipated it would require two and one-half to three years to complete the project. Other companies

General Petroleum Corporation's fireflood on the Marina lease at South Belridge utilized a five-well pattern consisting of four producing wells, marked by pump units, and an injection well in the center, marked by a standard steel derrick. (Mobil Oil Corporation)

sharing the estimated cost of $1 million in return for being privy to the results of the experiment included Continental Oil Company, Esso Research & Engineering Company, Gulf Oil Corporation, Honolulu Oil Corporation, The Ohio Oil Company, Shell Development Company, Sunray Mid-Continent Oil Company, The Texas Company, Tidewater Oil Company and Union Oil Company of California.

Field work for the Midway-Sunset project also had begun in the summer of 1955, with California Research handling the project in cooperation with the Western Division of Standard Oil Company of California, the parent firm. The technique was one that had been developed at the research company's La Habra laboratories.

The site of the experiment was on Standard fee property. The initial preparation had consisted of drilling four wells, including an injection well and three producing wells which encircled the injection well at distances of some 200 feet. It was anticipated that the project would cost approximately $1 million. Company engineers expressed confidence that the experiment would produce more oil. They said the question was whether it would prove economically feasible.

Inevitably, there was some public apprehension about the new technique. Some people conjured up visions of a raging fire leaping out of the ground to engulf bystanders. Hank Ramey, who helped direct the South Belridge burn, recalled that some concern was even expressed that the quantity of air being compressed for injection into the ground would result in a permanent low-pressure area, which in turn would cause perpetual storms.

As the firefloods proceeded without undue incidents, fears subsided, and engineers settled down to the task of monitoring the phenomena they had created in the earth. If they had been detectives, they would not have had to sift clues any more carefully than they did in keeping tabs on the unseen infernos.

The temperature of the oil that came out of the ground was one clue. It indicated how close the combustion front was to the producing well. The composition of the gases

Phil Witte with bottomhole heater used to heat two shallow wells on Joseph McDonald Oil Company's Shale lease in the Midway-Sunset field. (William Rintoul)

from below was another clue. The air injected into the oil reservoirs was, like all air, composed of 21 parts oxygen and 79 parts nitrogen. If oxygen persisted in the produced gases, the underground combustion was obviously inefficient. On the basis of such clues, engineers had to decide the rate at which air should be pumped downward. If the air injection was too slow, the combustion front would come to a virtual standstill. If the injection was too fast, the burning profile became distorted.

While firefloooders worked to perfect their technique, producers turned in increasing numbers to another tool that offered an approach to the application of heat to heavy crude.

The tool was the bottomhole heater, a device to apply heat down the hole opposite the producing sand. The system consisted of two separate units: a surface unit which was called the heater and a subsurface unit known as the heat exchanger. The surface unit came completely assembled on a metal base; the subsurface unit was

Tidewater Oil Company engineers
Ernie Young, left, and Mike Bealessio
with one of the bottomhole heaters used
to boost production in the Kern River
field. (William Rintoul)

assembled while tubing was being run into the well. Controls were on the surface unit; they were automatically regulated by the pressure of the circulating pump and the temperature of the circulating fluid, which could be water or oil. The fluid normally went into the hole at temperatures up to about 300 degrees and came back at about 200 degrees, making the trip to and from the heat exchanger in separate strings of small-diameter pipe run with the tubing in a single operation. The heat exchanger applied heat to the producing formation by radiation.

A strong selling point for the bottomhole heater was the system's relatively reasonable cost. The $1 million outlay that seemed to be the entry fee for a go at the still-to-be-proved fireflooding technique effectively limited the ranks of the fireflooders to major companies. On the other hand, for an investment of around $3,000, anyone could put a bottomhole heater to work, and if the heater did not do the job on a particular well, there was always the chance it might to better at another.

On the 50-acre Shale lease on Sec. 26, 31S-22E, Kern County, near Fellows in the Midway-Sunset field, Joseph McDonald Oil Company installed four heaters to service five wells. Four of the wells had been drilled in 1914, the

Not a man from Mars, but a steam operator dressed in a safety suit. The aluminum helmet had a mirrored, one-way eyepiece to reduce heat. Richard Penny, Union Oil Company of California, modeled the safety suit in the Guadalupe field. (Union Oil Company of California)

fifth in 1941. Production was from the Potter sand with depths ranging from 800 to 1,400 feet and gravity from 12 to 14 degrees.

The bottomhole heaters doubled production from the 27 barrels per day the wells were pumping before heaters to 54 barrels daily after heaters had been installed. The average cost of the heaters, including installation, was $3,000 each, figuring out to $2,400 per well. The payout time varied from well to well, depending on the increase in production. With the price of crude pegged at $1.94 per barrel, payout ran from four months for one well to as long as two years for another.

A collateral benefit was that the oil reached the surface warmer than before, making the crude easier to handle, especially in the winter.

In the Yorba Linda field in Orange County, Western Gulf Oil Company credited bottomhole heaters with an average increase in production of eight barrels per day for 17 wells that produced 12.8-gravity oil from sand encountered at about 600 feet. The company said increased production paid out heater installations in less than three months.

The most successful use of bottomhole heaters came in the Kern River field, where Tidewater Oil Company over a one and one-half year period in 1957-1958 installed heaters in 54 wells, increasing production from 250 barrels a day to 1,443 barrels daily, an increase of almost 600 percent. The average production increase per well was 22 barrels per day, substantially better than the average of 10 barrels daily generally credited to bottomhole heaters for other installations in California. One well on the San Joaquin property went from 2 barrels daily to 52 barrels per day. The use of heaters by Tidewater and other operators was largely credited with increasing Kern River's production by some 2,000 barrels a day over a two-year period, from 12,390 barrels daily in 1956 to a high of 14,440 barrels per day in 1958.

The increases in production from bottomhole heaters were welcome, of course, but to engineers, it was obvious that the heat radiating out from heat exchangers penetrated only shallowly into the oil sand, representing something akin to an effort to heat a cavernous room with matches. For better recovery, it would be necessary to deliver heat deeper into the sand.

Meanwhile the results were beginning to come in from the experiments with fireflooding. In the Midway-Sunset field, California Research Corporation discontinued its burn in May 1957, some 10 months after initiating combustion. The company did not release details, but the consensus among industry observers was that the project had not been commercially encouraging.

From South Belridge, General Petroleum Corporation checked in with a more positive, though qualified, report. The company confirmed that fireflooding did indeed increase recovery. Production from wells involved in the experiment on the Marina lease had been increased from 20 barrels per day before ignition to a high of 390 barrels per day, averaging about 140 barrels daily during the test, or seven times the rate before ignition.

With normal methods of production, the company said, a recovery of 10 to 15 percent of the oil in place might have been expected during the life of the field, estimated at 40 to 60 years. With fireflooding, recovery represented 51 percent over a period of one and one-half years.

Unfortunately, the cost per barrel of oil recovered was far higher for fireflooding than for normal production. The company said it had been necessary to rework wells more often, including scratching perforations, swabbing, bailing and injecting hot oil into the casing. Rapid corrosion and high temperatures had caused the failure of conventional steel pipe and equipment, both above and below the surface, creating costly maintenance problems.

The jury, in effect, was still out.

In April, 1960, Shell Oil Company quietly began testing another approach in the Yorba Linda field, injecting steam into a shallow zone, known as the Upper Conglomerate, that produced 12-gravity oil from a depth of about 600 feet. The company had tested the steam-heat method in

Two Halliburton portable steam generating units on a job for Standard Oil Company of California in the Kern River field, June, 1964. (William Rintoul)

the laboratory, and the results had been encouraging. The crude with only a relatively moderate increase in temperature showed a dramatic decrease in viscosity, which, of course, meant it would flow more readily. The company said the cost of the test might be as much as $150,000, and that it might be a considerable period of time before results were known. If the pilot program proved successful, the company said, a full-scale operation might be attempted, but such an operation would be at least several years in the future.

Steam was hardly new to oil fields. Drillers had used it for many years to power the rigs with which they drilled wells. By the 1940s, they had largely relegated it to the discard heap in favor of powering rigs with gas and diesel fuels. Pipeliners also knew about steam. They had used it for years to heat heavy crude so that the oil might be pushed more easily through pipelines.

The suggestion had even been made as early as 1901 by a man named J.W. Goff, on a visit to Bakersfield from his home in San Diego, that those having trouble producing the heavy tarlike oil in the nearby Kern River field might profitably give some thought to injecting steam into wells to heat the oil. Goff had obtained permission to experiment with steam in a well on the Golden Rod property. He had run a steamline into the well. When the crude did not respond to steam as well as Goff had hoped it might, he ran air into the hole, also without pronounced success. Then he rigged the air line to run through the steam line so that air would be hot when it was injected into the well. The well flowed steadily for a week, but with the price of oil

Steam generator capable of producing 18½ million Btus of steam per hour was lifted off a low-boy trailer to given an early boost to Tidewater Oil Company's steaming operations in the Kern River Field. (Getty Oil Company)

declining, the cost of the operation exceeded the benefits and the experiment came to an end.

In the months that followed the first injection of steam at Yorba Linda, there was no word from Shell to indicate whether the pilot project was a success, or even if it showed promise.

Before the year ended, Shell had begun preparations to inject steam in the Coalinga field. The company made no early announcement of its plans and, in fact, seemed disturbed when news leaked out. The zone selected for the pilot test was known as the 1st zone. It was an isolated sand in the upper portion of the Temblor zone that produced 15-gravity oil from a depth of about 1,000 feet. The program called for the drilling of an injection well on Sec. 29, 19S-15E, Fresno County, around which were to be grouped four observation wells at a radius of 90 feet from the injector and two producing wells at a distance of 180 feet and on opposite sides of the injector. The company planned to use the observation wells for recording temperatures, and to run frequent thermal surveys in order to detect any possible migration of steam from the oil zone.

Interestingly, the injection well was only one-quarter mile east of what formerly had been the Section 30 Oil Company property. Shell had purchased the property only a few months before from Lyle and Walter Fisher. Production dated back to 1906, when the property had been controlled by Balfour, Guthrie & Company. There were 47 wells on the 520-acre spread, and they produced about 240 barrels daily, or about 5 barrels per day per well. Shell reportedly had paid approximately $1 million, or roughly $4,200 per barrel of daily production, as such acquisitions were commonly expressed.

In March, 1961, Shell began putting away steam. Though the company had nothing to say about results, the continued steaming at Yorba Linda and the expansion of steaming into the Coalinga field were not lost on competitors.

In the Kern River field, Tidewater Oil Company, though it had enjoyed more success than any other California

operator with bottomhole heaters, entertained no illusions that such an approach would be an ultimate, economic answer for the large-scale application of heat.

The company had experimented with thermal methods in California as early as 1923, when engineers injected hot water into wells in the Casmalia field eight miles south of Santa Maria. The field produced heavy crude from depths as shallow as 700 feet. The early work had produced some information, but the equipment that was available, the limited development at the time of oil reservoir engineering principles and, more importantly, the availability of prolific light-oil fields, which were much easier to produce, plus the technical problems of running heavy oil in refineries of the day, had not encouraged further work in thermal recovery.

Initially, Tidewater's engineers favored hot water over steam, reasoning that if a limited amount of hot water in closed pipes in the bottom of a well, that is, a bottomhole heater, would bring a response in heavy oil production, a better response could be obtained by using a lot more hot water, either pumping it directly into the producing well, or better yet, as a standard waterflood, using the hot water both to heat the oil and to push it through the reservoir to producing wells.

Laboratory work proved so encouraging that the company initiated a two-well hot-water injection test at Kern River in August, 1961. This was followed a year later by a more ambitious heavy oil recovery stimulation experiment, code named Project HORSE. The highly secret project involved five wells on fee property on Sec. 32, 28S-28E, Kern County. The company began injecting water in August, 1962, putting away the 300-degree water into four corner wells at a rate of about 2,000 barrels a day. Over a period of some eight months, the company injected almost two million barrels of hot water, increasing production from the center well to as much as 100 barrels a day of oil, finally discontinuing water injection when tracer tests indicated severe channeling.

While Project HORSE was under way, Tidewater also began experimenting with steam, injecting the steam into wells on an individual basis, allowing the well to soak and then returning it to production, at an increased rate, it hoped.

Whatever the differences in the Tidewater and Shell approaches, both companies shared one concern. For competitive reasons, both operated with tight security. Visitors were turned away from project sites; inquiries were deflected with no release of information.

Concurrently, there was another interesting development in California's heavy crude fields, particularly those in the San Joaquin Valley. Almost overnight the producing leases owned by independents, especially stripper leases, became hotly sought-after items in what before the end of 1962 came to be described as the greatest seller's market in the history of California oil.

The buying spree in one field alone—the Kern River field—resulted in more than 300 wells changing hands in a 10-month period. Going prices rose to figures that were incomprehensible to many observers. Some sales brought prices as high as $9,000 per barrel of daily production. For example, if the property happened to be producing 20 barrels a day, the price would be $180,000, an unheard of figure.

The prices made ancient history out of those of five years earlier, when $2,000 per daily barrel was considered par for the course, and even dwarfed those of one year before, when $3,500 to $4,500 per daily barrel sufficed to transfer producing properties from one owner to another.

What lay behind the sudden sky-high prices? The answer certainly did not lie with any increase in the price paid for Kern River crude oil. On the contrary, the barrel of Kern River crude that brought $2.55 five years before only brought $1.75 in 1962—a decrease of more than 30 percent.

Significantly, the two companies that were leaders in lease acquisitions at Kern River—Shell Oil Company and Tidewater Oil Company—happened to be the leaders in experimenting with steam and hot-water recovery techniques.

Shell in less than a year's time purchased no fewer than 14 producing leases from six independent operators at Kern River, acquiring more than 200 wells. Among those who sold to Shell were Nate Morrison, D.D. Lucas, E.A. Clampitt Oil Company of Oildale, P.D. Mitchell, Producers Oil Corporation of America and C.E. Rubbert.

Tidewater purchased nine producing leases from five operators at Kern River, acquiring more than 70 wells. The purchases were made from Ventura Oil Company, West Crude Oil Company, D & D Oil Company, Butler-Richardson and Producing Properties Inc.

While the pace of lease purchases was hotter at Kern River than elsewhere, the same pattern was being repeated in other fields. Shell's purchases in other San Joaquin Valley fields during the year involved more than 300 wells in addition to those picked up at Kern River. The acquisitions included the Ralph R. Whitehall and McGreghar Land Company wells at Mount Poso; Pacific

Coast Gasoline Company and Anchor Oil Company leases at Midway-Sunset; Maxwell Hardware Company, Oil Division, leases at Round Mountain; Producers Oil Corporation of America leases at Coalinga; and Producing Properties Inc. leases in the Edison, Mountain View, Mount Poso, Round Mountain and Coalinga fields.

Though details were lacking, it was obvious there had been a breakthrough in thermal recovery techniques that would add value not previously present.

Overnight, almost everyone scrambled to get in on the steam act. Fortunately, the economics looked good, at least for the short run. Unlike fireflooding, the entry fee was not so large that it automatically kept out the small operator. Steam was within reach of most producers. And it applied the heat in a hurry to get an almost immediate return. An operator might invest $30,000 for a steam project and, with luck, make enough in six months to pay off the investment.

Steaming easily won out over fireflooding in a comparison of the investment required. A steam generator that would convert 1,200 barrels per day of water into steam cost $35,000 to $45,000. A smaller generator, one capable of converting 500 barrels per day of water into steam, could be had for around $20,000. For fireflooding, an operator faced an outlay of more than $100,000 for a compressor capable of furnishing air to sustain underground combustion. And the return would not be so fast. It took time for the heat to make itself felt in increased production.

Another factor put steam within the reach of even the small producer. The producer could rent the equipment for a pilot test before making any substantial investment in steam generators. One service company offered a portable steamer that would inject approximately 650 barrels of water converted to steam at 10,000 Btus per hour. The unit featured an injection pressure of 700 pounds-per-square-inch and injection temperatures from 425 to 500 degrees Fahrenheit. If the service company furnished the men and fuel to operate the unit, the cost was $262.50 for the first four hours, $50 an hour afterward. If the producer furnished the fuel and the service company furnished the men, the cost was $262.50 for the first four hours, $25 an hour afterward. If the unit was furnished on a leased basis, for a minimum of 10 days the cost was $1,250 for the first day and $125 per day afterward. On a monthly basis, the charge was $1,800 per month, $75 per day after the first month.

Another manufacturer of steam generators rented generators for approximately $1,200 per month with the oil operator supplying both the men and the fuel.

In short order, operators were standing in line for units. The waiting lists ran three months or longer. Steam fever brought boom times to makers of steam-generating units and water treatment equipment. Companies like Struthers Wells Corporation, Baker Oil Tools, Texsteam, Halliburton Company and National Tank Company participated in a seller's market. One salesman in four months reportedly sold more than $1 million worth of steam generators.

In light of the overnight manner in which steaming had burst on the California scene, the competitiveness of the

Occidental Petroleum Corporation joined the ranks of the steamers on the Amber lease in the Midway-Sunset field, using a National Tank Company unit capable of thermal output of 25 million Btus per hour to inject steam into the Potter sand at 1,600 feet. (William Rintoul)

situation and the rapid development of the technique, most operators chose to maintain a deep silence about steam projects. Most were reluctant even to acknowledge projects, much less offer any information on results.

Scouts who traditionally had watched exploratory wells for the clues that would indicate an oil discovery suddenly were traveling dusty oil field roads looking for steam generators, trying to put together a picture of who was steaming and whether they were doing any good. Operators responded with "Keep Out" signs and locked, sometimes guarded, gates. One major company went so far as to make it required operating procedure that paper bags be placed over all gauges and removed only when authorized personnel found it necessary to read the gauges, after which the bags were to be replaced.

Rumors flew, with each more exciting than the last. There were unsubstantiated reports of fantastic increases in production, of wells that had been all but ready for abandonment surging back after steaming to behave like veritable gushers.

One heard, for example, that out at South Belridge the field's biggest producer, Belridge Oil Company, had injected steam for 15 days into a well that had previously pumped 27 barrels a day. The well was returned to production making more than 250 barrels daily.

As Casmalia, Union Oil Company of California was reported to have injected steam into a well that had netted 14 barrels a day of oil. Afterward, the well averaged 150 barrels a day net for several days, finally settling to 50 barrels daily.

At Cymric, several miles south of the South Belridge field, a Standard Oil Company of California well was rumored to have jumped from 30 barrels a day to 300 barrels a day, thanks to treatment with steam.

At Guadalupe, a heavy crude field 11 miles west of Santa Maria, a Union well that had been producing 30 barrels a day from the Sisquoc sands at about 2,800 feet was said to have come back after two weeks of steam injection and three days of soaking to produce 300 barrels a day. Two

months later, the well reportedly was continuing to produce 250 barrels daily.

At Kern River, Tidewater Oil Company was rumored to have increased production from an undisclosed number of shallow wells from a presteaming 100 barrels daily to a poststeaming rate of 430 barrels a day.

At Midway-Sunset, Mobil Oil Company, the successor to General Petroleum Corporation, steamed a shallow well and reportedly boosted production from 10 barrels a day to more than 100 barrels a day.

Initially, steaming was largely on a trial-and-error basis, with the most popular method what was first known as push-pull and later came to be commonly called huff-and-puff. The operator injected steam into a given well for a particular number of days, perhaps 10 days, let the well sit, or soak, for perhaps another 10 days, then returned the well to production.

A logical extension was the steam-drive, or steam-flood, in which steam was injected continuously into an injection well to heat and push oil toward producing wells through which the heated crude might be recovered.

With the spread of steam projects, a share of the secrecy surrounding the work began to fall away. The technique that had had its inception in Shell Oil Company's small pilot operation in the Yorba Linda field became everybody's business, and producers began to confirm the production increases that had only been rumors before.

On a summer day in 1964 in the North Midway portion of the Midway-Sunset field, an employee of Occidental Petroleum Corporation erected a sign beside an oil field road three miles northwest of the town of Fellows.

The sign read: "Danger. Steamflood in Progress. Enter at Your Own Risk."

The sign marked the access road to the company's Amber lease. It was, as far as anyone knew, the first acknowledgement by any oil company to the general public of the use of the recovery method that would make it possible to produce billions of barrels of additional oil from California fields.

11 The Man Who Was Right

The check from Richfield Oil Corporation, Los Angeles, California, to John P. Louden, Fairfield, Iowa, was dated May 26, 1958. It represented payment for the first year of a five-year oil and gas lease taken by the oil company on an 80-acre parcel owned in equal parts by Louden and his aunt, Mrs. Antoinette McMillen Louden, who also lived in Fairfield. The parcel comprised the east half of the southwest quarter of Sec. 34, 12N-22W, Kern County, California. It was located 20 miles southwest of Bakersfield in what was known locally as the Maricopa Flats area. Under terms of the lease, Richfield, as lessee, was given the right to drill for oil and gas. If the company discovered production, it would be entitled to a five-sixths share in return for having financed the drilling operation, reserving to the Loudens, as lessors, a one-sixth interest, equally split between John P. Louden and his aunt. The oil company's payment for the lease was at a rate of $10 per acre for each year of the specified five, or a total of $800 for the first year's rental. The check to Louden, like the check that went to his aunt, was drawn for the amount of $400.

In the overall scheme of things, Richfield's acquisition of the Louden lease was a routine leasing transaction, one of thousands entered into each year in California by oil companies putting together the acreage blocks that are a necessary prelude to drilling for oil and gas. To the oil company, the lease was one more to be marked on a map, recorded and filed safely away against the day that an exploratory well might make a discovery on the block of which the lease was a part. If the well found oil or gas, as had happened in the case of the huge leasehold taken a decade before by the same company in Cuyama Valley, the Louden lease conceivably could be worth millions of dollars, depending on where it was with relation to the discovery. If the wildcat proved to be a dry hole, the shares of interest divided between Richfield Oil Corporation, John P. Louden and his aunt would be of no value. The company eventually would

John P. Louden. (Helen Louden)

quitclaim the lease, turning it back as a footnote to one more geologist's dream that had gone astray.

If the taking of the Louden lease was a run-of-the-mill occurrence for Richfield, for Louden it was anything but routine. It was the opening of a door into a world about which he knew little and one in which he previously had not had reason to think of himself as a participant, the world of oil exploration. At 50, he had been retired four years from the Louden Machinery Company, which his great uncle, William Louden, had started in Fairfield in 1867. The plant manufactured what its advertising brochures described as the Louden Line of Barn Hardware and Equipment, including stalls and stanchions, litter and feed cars, hay tools, ventilation equipment, horse stable equipment, overhead door equipment, water bowls, feed trucks, steel pens, windows, sliding-door tracks and hog house equipment. The town of Fairfield was a pleasant one of some 8,000 people in the southeast corner of Iowa. It was a trading center for the area's hog-and-grain farmers and enjoyed a degree of stability and prosperity through the presence of various manufacturing plants, being the home not only of the Louden plant but also of Philco Washers and Iowa Malleable Iron Company. It was also the seat of Jefferson County, conducting the County's business out of a splendid courthouse that had been built in 1891, and the site of Parsons College, a private institution of higher education. The town enjoyed a life style far removed from the boom-or-bust atmosphere that accompanies drilling for oil.

The lease taken by Richfield Oil Corporation offered an entry for Louden into the fantasy that must surely lie in the mind of virtually every landowner in the United States whose parcel of ground is large enough to accommodate the equipment necessary for the drilling of a wildcat well, the fantasy that someday someone may come along and discover oil or gas under the land, bringing with the discovery wealth and excitement far beyond the monotony of day-to-day life. The lease indicated in tangible terms that at least one oil company, and that company one of the

On a trip to Bakersfield to check on the land he was convinced held oil, John Louden with his wife, Helen, right, and sister-in-law, Winnie, visited the site of the Kern River discovery well which had started the Kern County oil industry on its way in 1899. (William Rintoul)

industry's most successful wildcatters, thought there might be oil or gas under the Louden property. It was a clarion call to John P. Louden to get acquainted with a whole new world, one to which he might bring little in the way of knowledge but much in the way of enthusiasm. If he knew little about what prompted companies to drill exploratory wells in a given area, either of the economics or geology involved, he brought no lack of another ingredient necessary to the success of oil ventures. He brought a willingness to believe.

The parcel of land in Kern County, California, had been owned by Loudens through three generations. Thomas Louden, John's grandfather, had purchased the raw land by mail in the 1880s, apparently for no more reason than a strong faith in the ownership of land and the feeling that California's San Joaquin Valley was as good a place as any in which to invest for the future. The land at that time had never felt the plow, nor would it be cultivated for almost seventy years more. There had been some minor oil activity in California, but nothing of importance, and oil was still some twenty years short of being an economic factor in the state.

In the 1920s, Thomas Louden handed on the land to his two sons, Walter, who was John's father, and Robert Roy Louden, with the condition that they were to assume the obligation of paying the taxes assessed against the property. Though there must have been times when the two sons wondered about the wisdom of paying taxes on distant land that produced no income, particularly through the years of the Depression, they regularly mailed in the assessed payments to Kern County Tax Collector Perry Brite.

Once John's father traveled to California to see the property. While driving through the community of Greenfield a few miles south of Bakersfield, he passed a real estate office with a sign which listed the realtor as Roy Loudon. Though the name was spelled differently, he stopped to see if there might be a connection and had an interesting talk with Roy Loudon and his wife Anne. The two men were unable to trace a relationship, but it seemed apparent they came from the same family tree, considering physical similarities and the fact that Roy Loudon, though born in Kansas, had lived in Iowa.

With the deaths of Walter and Robert Roy Louden in the early 1950s, the parcel of Kern County land passed on to their heirs, one share to John Louden, the other share first to Antoinette McMillen Louden, Robert Roy Louden's widow, and subsequently to her son Tom, John's cousin. John Louden traveled to California to see the land, and renewed acquaintance with Roy and Anne Loudon.

In 1953, the Iowa Loudens leased the surface of the Kern County parcel to Robert Pelletier, a Bakersfield farmer. Pelletier, like Richfield, paid a rental of $10 a year per acre, buying with the payment the right to grow cotton on the

Roustabouts hastened to hook up lines in preparation for the completion of Richfield Oil Corporation's KCL H No. 34-9, the wildcat that culminated a more than 20 years' search with what shaped up as a significant oil discovery at San Emidio Ranch. (William Rintoul)

land. In effect, when Richfield leased the oil and gas rights in 1958, it was like doubling the payment from Pelletier, though the oil company, of course, had plans other than the growing of cotton.

On a warm July morning some six weeks after Richfield Oil Corporation had leased the Louden parcel, the company had the happy occasion to do something which it, among California oil companies, probably did best. Richfield gathered together in Los Angeles a group of upward to 50 persons, including members of the oil and financial press and officers and department heads of the company, loaded the group aboard two chartered buses and headed north on Highway 99 for a 110-mile trip into the San Joaquin Valley. Once in the valley the buses turned west on Highway 166, also known as the Maricopa Flats highway, and drove for a distance of nine miles, turning off the highway to stop beside a steel derrick that marked the site of Richfield's KCL H No. 34-9, a newly drilled exploratory well on Sec. 9, 11N-21W, Kern County. Two other, smaller groups also converged on the well, including one made up of officials from Kern County Land Company's San Francisco headquarters and another that included key personnel from Richfield's Northern Division office in Bakersfield.

Among the financial writers present were Howard Kegley, the dean of California's oil columnists, who had been oil editor of the *Los Angeles Times* for some 30 years before retiring to syndicate an oil news column which was published in the *San Francisco Chronicle, Long Beach Press-Telegram, Fresno, Modesto* and *Sacramento Bees, Salinas Californian, Santa Barbara News-Press, Santa Cruz Sentinel-News* and *Santa Paula Chronicle;* Bob Sullivan, oil editor of the *Los Angeles Times;* Carl Lawrence, West Coast editor for *Oil & Gas Journal,* a weekly magazine published in Tulsa, Oklahoma, and widely known as the Bible of the oil industry; Russ Palmer, editor of *Petroleum World and Oil,* a weekly trade journal of the oil industry published in Los Angeles; and Bill Gregg, West Coast editor of *The Oil Daily,* an industry tabloid published in Chicago, Illinois.

Richfield gave newsmen a list of company officials present. The list was democratically arranged in alphabet fashion and included from the company's home office N.F. Anderson, manager, pipe lines; M.L. Arnold, manager, gas operations; W.T. Autrey, comptroller; E.M. Benson, manager, joint and foreign production operations; Cleve B. Bonner, treasurer; C.W. Coughlin, assistant to the comptroller; David E. Day, vice president and manager of manufacturing and transportation; Rollin Eckis, executive vice president; J.T. Foster, general sales manager, retail and operations; Marvin L. Gosney, director; Colin W. Graves, press relations representative; Mason L. Hill, manager, exploration division; H.T. Hutchinson, general sales manager, wholesale; Charles S. Jones, president; W.K. Kreiger, manager, operating division, production department; Frank T. Lloyd, chief production engineer; J.W. Mathews,

Watching the first spray of oil flow from the San Emidio Nose discovery well were officials of Richfield Oil Corporation, backer of the well; Kern County Land Company, owner of the land on which the discovery was made; and a host of newsmen on hand for completion of Richfield's KCL H No. 34-9. (William Rintoul)

234

chief geophysicist; William H. McCloud, chief special agent; C.R. McKay, manager, land and lease department; William A. Miller, manager, insurance department; M.L. Natland, regional geologist; D.F. Purdy, assistant manager, pipe lines; R.W. Ragland, vice president and manager, legal department; R.W. Rood, assistant to the president; I.T. Schwade, regional geologist; Norman F. Simmonds, secretary; E.V. Squires, director of purchases; Stender Sweeney, vice president and manager, industrial relations; W.F. Tegtmeier, staff consultant; Frank B. Tolman, geologist; and W.J. Travers, vice president and manager, production.

Those present from Richfield's Northern Division included George Austin, Northern Division landman; M.C. Barnard, Northern Division scout; I.M. Bumgardner, San Joaquin Valley District production superintendent; Stan A. Carlson, Northern Division paleontologist; C.K. Curry, manager, Northern Division production; Lee V. Gefvert, production engineer; Clarence E. Gotterba, superintendent, gas department; Hobart A. Harvey, pipe line superintendent; George R. La Perle, geologist; C.B. Lisman, assistant production superintendent, Cuyama Valley; Gene Nichols, production engineer; Ray E. Pearson, Northern Division geologist; Deryl E. Rising, general drilling superintendent; Joe P. Shea, development engineer; H.G. Smith, chief clerk; Warren G. Stoddard, geologist; W.S. Tanner, Cuyama district superintendent; R.J. Wells, Northern Division geophysicist; and R.G. Wilfert, San Joaquin Valley District superintendent.

From Kern County Land Company there were John T. Pigott, chairman of the board; George G. Montgomery, president; Crawford Greene, director; Porter Sesnon, director; Herbert L. Reid, executive vice president; John H. Matkin, vice president; Carter H. Breusing, vice president; George L. Henderson, vice president; Lee M. Ralph, manager, oil division; Ralph E. Brodek, chief geologist; Robert Horton, geologist; Chester Eaton, petroleum engineer; E.K. Parks, consulting petroleum engineer; and James A. Walker, manager, land.

At the well, Richfield had a prepared release for newsmen. The first paragraph read:

"Richfield Oil Corporation today completed its KCL H 34-9 test well in the southern end of the San Joaquin Valley, 21 miles south of Bakersfield, marking the discovery of a new oil field to be named the San Emidio Nose field. The well was brought in flowing at a rate of _____ barrels per day of _____ gravity oil through a _____ inch bean with _____ MCF rate of gas, _____ tubing pressure."

After visitors had had a reasonable amount of time to look at the drilling rig and renew acquaintances with each other, the group assembled by an earthern sump, standing safely back from a pipeline that ran from the wellhead into

Left to right, Charles S. Jones, president, Richfield Oil Corporation and George G. Montgomery, president, Kern County Land Company, happily discussed completion of Richfield's San Emidio Nose discovery well on a large block leased from Kern County Land Company. (William Rintoul)

the sump. The crew on Richfield's Rig. No. 10 turned the well to the sump. A dark spray of oil roared into the sump, shaking the pipe through which it flowed.

By this time it was nearly noon.

Charles Jones, Richfield's president, said the company would gauge the well and have a production rate in another hour or so.

Bill Clark, Richfield's photographer, got stung by a bee, suffering a reaction which prevented him from taking pictures.

The buses revved up, and the group boarded for the short ride south over the flat valley floor to a cluster of buildings at the foot of the Coast Range that formed the boundary of the San Joaquin Valley. The buildings were the headquarters for Kern County Land Company's San Emidio Ranch, on which the discovery well had just been completed. At the ranch, the land company and Richfield hosted a barbecue.

Afterward, there were several short speeches.

Charles Jones, though an obvious and willing participant in the promotion involving completion of the discovery well, officially adopted a conservative attitude with regard to the significance of the find. "One well doesn't make an oil field," Jones said, "or one swallow a summer."

He indicated that more wells would have to be drilled before the extent and importance of the discovery could be determined.

On a less cautious note, he added that it might be necessary for Richfield to enlarge the capacity of its main pipeline system, fortuitously located less than one-quarter mile from the discovery well. The pipeline system carried San Joaquin and Cuyama oil to the company's Watson, California, refinery, 115 miles to the south. Jones said the increased capacity might be required as the new field matured. He said additional line facilities, if built, would also tie in with the nation's defense effort by providing potential capacity to carry oil from the Elk Hills field to Southern California refineries in event of a national emergency.

The discovery of the Yowlumne Field in January, 1974, brought drilling back into the area where the 80-acre Louden parcel was located. Loffland Brothers Company's Rig No. 34, center, drilled the discovery well. (William Rintoul)

Mason Hill, Richfield's exploration manager, briefly traced the history of exploration in the area, stating that the San Emidio Nose discovery culminated an intense subsurface geological study and test well program which had been carried on in the area by Richfield and others almost continually since 1935.

One of the first companies to show interest in the San Emidio area, Hill said, was The Ohio Oil Company, which had made a seismic survey in the 1930s and in 1935 drilled an approximately 7,000-foot duster. The next company to tackle the play was The Texas Company, which deepened the Ohio duster to about 10,000 feet without finding pay sand. Then Shell Oil Company came to bat, drilling a deep duster in 1939.

Richfield, Hill said, had drilled its first wildcat at San Emidio in 1945, going to 12,674 feet, which was the deepest well the company had drilled up to that time. Continental Oil Company moved in next, deepening the original

Two who played important roles at Yowlumne were, left to right, Jim Dorman, who succeeded Don Taylor as Tenneco's California exploration manager, and Lane Howell, Tenneco's geophysical manager. (William Rintoul)

Ohio duster some 400 feet. Later Continental drilled a new hole and again failed to hit. In 1951, Western Gulf Oil Company drilled a 12,172-foot duster which, like some of the earlier tests, found oil-stained sand but not enough permeability to make a field. Richfield reopened the play in 1957, drilling a 14,471-foot duster southeast of the discovery well. The company moved up structure to drill the discovery well, which together with the earlier two Richfield tests added up to a total expenditure for the company of about $1.2 million in San Emidio exploration.

In reviewing the dry holes that had been drilled before the discovery, Hill said they illustrated, "about as well as anything could," why the oil industry needed the 27.5 percent depletion allowance. "Our backs are to the wall," Hill said. "We need help from everybody."

Before the luncheon ended, Bill Travers, vice president and manager of production, announced that the initial rate of production for the discovery well was 2,520 barrels per day net of 29.8-gravity oil, based on a one-hour gauge. The flow was through a 1½-inch bean with 740,000 cubic feet per day of gas and 100 pounds-per-square-inch tubing pressure.

Newsmen filled in the blanks on the release they had been given earlier.

The release, in addition to providing a lead paragraph, went on to state that the well was producing from approximately 100 feet of oil sand in the interval 11,452 feet to 11,560 feet and that production was from the Reef Ridge sand of Upper Miocene age.

One paragraph read: "The San Emidio Nose discovery well is located on a lease comprising 8,500 acres of land owned by the Kern County Land Company. In addition, Richfield holds 58 individual leases covering 3,700 acres in the area of prime interest."

In Fairfield, Iowa, John P. Louden, whose parcel with its 80 acres represented approximately 0.66 percent of Richfield's total lease block of 12,200 acres, responded to the news of the oil discovery in California with understandable elation and excitement.

This was the news he had been waiting for. True, the discovery was not on his land, but it was on the acreage that Richfield had leased to look for oil, and it must surely bode well for the rest of the land the company had acquired, including the Louden parcel.

Louden, of course, had not been present at the barbecue at Kern County Land Company's San Emidio Ranch and hence had not been privileged to hear Mason Hill's brief but illuminating, one might say sobering, account of the unsuccessful drilling that had preceded the discovery of the San Emidio Nose field. He had no particular frame of reference in which to place the discovery, either in light of the dry holes that had been drilled before or the dusters that might be drilled later. Nor was there any evidence that if he had more fully understood the formidable odds that accompanied drilling for oil he would have been any less enthusiastic.

One of the first things that Louden did was to locate on a map the discovery well with reference to his property. The well was five miles to the southeast. To an experienced oilman knowledgeable about the realities of oil field geology, the five-mile distance would have raised an immediate

Drilling rigs worked around the clock to develop the Yowlumne field. Above, left to right, Loffland Brothers Company's Rig No. 185 on location at Tenneco's Tenneco-Texaco-Smith No. 65X, Loffland's Rig No. 92 on location at Tenneco's Tenneco-Texaco-McMahon No. 67X, Loffland's Rig No. 16 on location at Tenneco's Tenneco-Texaco-Yowlumne No. 45X, Montgomery Drilling Company's Rig No. 4 on location at Chevron's No. 46-33 and Loffland's Rig No. 34 on location at Chevron's No. 16X. (William Rintoul)

warning flag, cautioning against unbridled optimism. For Louden, accustomed to thinking of distance in terms of the flat mid-American prairie, five miles did not sound like a long way at all. It sounded longer when he learned that Richfield, for the first follow-up well, had moved only one-half mile northwest of the discovery and seemed to consider the second well a wildcat.

Fascinated by the world of oil that had suddenly come to his door in Fairfield, Iowa, more than 1,500 miles removed from the new-found San Emidio Nose field, Louden embarked on a course to learn everything he could about what was going on in the discovery area in the southern San Joaquin Valley, particularly as it had any bearing on the prospect of bringing in an oil well on his property. He subscribed to *The Bakersfield Californian,* which published a daily column of oil news, and to Taft's *Daily Midway Driller,* which also published oil news on a frequent basis.

In reading about oil activity, Louden saw the names of oil producers, engineers and geologists mentioned from time to time in accounts of particular wells. He secured a Kern County telephone directory from Tom Jarvis, the telephone company's manager in Bakersfield; looked up the addresses of these men; and began writing to various of them, asking for information and opinions on what they thought of activity in his area and of his land. He unfailingly enclosed a stamped, self-addressed envelope with his letters. Many of those to whom he wrote courteously responded, and to many he sent gift packages of blue cheese from the Maytag Dairy Farms in Newton, Iowa, 75 miles northwest of Fairfield.

The pursuit of information led him to *Oil & Gas Journal, California Oil World* and *Western Oil & Refining,* and he subscribed to these publications. He saw an advertisement for a publication called the *California Petroleum Register,* which listed the names and addresses of those active in California oil, and he bought a copy, using it to secure more names and addresses with which to broaden his list of correspondents.

Among those who took the time to reply to his letters were such oilmen as George Almgren, who was associated with Hylton Drilling Company, a contract drilling firm, in Bakersfield; Ray Arnett, Richfield geologist; George Austin, Richfield landman in Bakersfield; E.A. Bender, an independent producer in Bakersfield; Bob Ferguson, an independent producer with production at Asphalto; Glenn Ferguson, one of Bakersfield's more successful independents; Ed Gribi, a consulting geologist with offices in King City; Phil McCormick, who was with Cavins Company, a service company, in Taft; Mike Rector, a consulting geologist with offices in Bakersfield; Bud Reid, executive vice president of Occidental Petroleum Corporation; Tom Roy, a geologist with The Ohio Oil Company in Bakersfield; and Hy Seiden, the consulting geologist who had authored the Asphalto play.

Louden expanded his list of correspondents to include businesmen and others not directly involved in oil exploration, among them such men as Orville L. Bandy, a professor in the Department of Geology at the University of Southern California; Henry Bowles, president of Buena Vista Farms, Inc., a firm with large landholdings in Kern County; Walter Kane, publisher of *The Bakesfield Californian;* Kern County Supervisors Floyd Ming and Vance Webb;

Bill Moore, president of Tejon Ranch Company, another Kern County landholder; Tex Newby, who headed Southern California Gas Company's Bakersfield office; Jim Radoumis, executive vice president of the Kern County Board of Trade; Warde Watson, a prominent Bakersfield realtor and developer; and Gene Winer, who was the mayor of Bakersfield.

Along the way, Louden wrote to California's Governor Edmund "Pat" Brown, who responded courteously, though failing to shed any light on the oil situation at San Emidio Nose. Louden sent the Governor a gift pack of blue cheese.

In the beginning, the conclusions Louden drew from the information sent him, and from his own willingness to believe there was oil beneath his property, were all encouraging.

His friend and possible relative, Roy Loudon, who with his wife held landowner's royalty interest in several wells in the Mount Poso field, 12 miles northeast of Bakersfield, wrote that on the wall of the office of a consulting petroleum engineer he had seen a map with lines drawn to indicate a trend approximately three miles wide that ran from Wheeler Ridge field in the south end of the San Joaquin Valley westward through the new-found San Emidio Nose field to the Buena Vista field near Taft, including in the trend the Louden acreage.

As heartening as this information was, it was not nearly as positive as another report Louden received. In the letters-to-the-editor section of *The Bakersfield Californian*, Louden came across letters written by a man named Wade J. Gessell who described himself as "The Old Dowser." Gessell lived in Weldon, California, a small community in the mountains 40 miles northeast of Bakersfield. He claimed to have the ability, through his dowsing method, of being able to locate oil and to tell, conversely, where oil would not be found. Gessell liked to write to the newspaper with predictions about whether given wells would find pay sand, and in more instances than not, he was correct. His batting average mainly was based on the fact that he almost always predicted the wells would be dry. As wildcatters knew, this was a statistically safe prediction. On predicting discoveries, he was less accurate, having no known successes. At any rate, Louden wrote to "The Old Dowser" and the upshot of the correspondence was that Gessell traveled to the Louden property to evaluate the oil possibilities of the wildcat ground.

Marlin-West Drilling Company Inc. made its debut as a California drilling contractor at Yowlumne. The company's Rig No. 11 drilled Tenneco's Tenneco et al., Bennett No. 84-32. (William Rintoul)

Gessell carried out what he called a "Dowsing Survey for Petroleum Mineral Oil" on the Louden property and forwarded a report to Louden. The handwritten report read:

"From a point 170 feet west of J.P. Louden Northeast corner a line drawn south to a point about 400 feet west of the Southeast corner outlines the East perimeter of Oil Pool which extends west across the entire west portion of the 80 acres, about 60 acres of Oil Pool. This pool extends an undetermined distance into adjoining property to the North, West and South. The Oil Sands will produce at approximately same depth as San Emidio Nose field. The bottom of the Louden field is much deeper than San Emidio Nose. The Richfield Oil Co. has lease on Louden Field and I hope they begin drilling at Early Date." The report was signed: "Dowser—Wade Gessell."

While Richfield Oil Corporation kept as many as five rigs busy with the development of the San Emidio Nose field, staying close in by completed wells, events were moving on other fronts. Richfield had not been the only company to lease wildcat acreage in the vicinity of the field. Other companies had jumped in to pick up open ground.

Four months after Richfield had completed the discovery well, Bandini Petroleum Company, Los Angeles, in a joint venture with Hamilton Dome Company, Tulsa, began rigging up to drill a wildcat two miles northwest of the Richfield well. In Fairfield, John Louden located the well on a map, saw it was moving the play closer to his property and, through newspapers and correspondents, closely followed the wildcat through the two months the Bell & Burden drilling crews drilled it to a total depth of 12,656 feet. The well ran low. Bandini attempted to get Richfield to take the hole deeper. The deal fell through, and the well was suspended.

It was June, 1959, almost one year after the discovery, before Louden read of another well that might affect his chances, and this was one that promised to bring the play almost to his property. The Ohio Oil Company moved in a Camay Drilling Company rig to drill KCL R. No. 21-2 on Sec. 2, 11N-22W, four miles northwest of the San Emidio

Nose field and only one mile southeast of the Louden property. Louden eagerly followed the progress of the Ohio wildcat. The hole was dry at 13,045 feet.

Though disappointed by the failure, Louden refused to give up. He kept up a flood tide of correspondence—and packages of blue cheese—to California. The volume of cheese was such that Mel Campbell, manager of Maytag Dairy Farms, wrote Louden a personal letter in which Campbell said, "Thank you J.P. for all the nice orders that you keep continually sending us. We like this."

The exploratory pace slowed at San Emidio Nose. The wildcats that were drilled were farther from the Louden property. It was like grasping at straws to think that they might influence drilling on the Louden parcel, though Louden contined to write often, to express unshakeable faith and to wonder why no one drilled on his 80 acres. There was a pair of wildcats by Texaco, successor to The Texas Company, both to the southwest of the San Emidio Nose field and five miles southeast of the Louden property. Both were dry. Rocky Mountain Drilling Company teamed up with Mohawk Petroleum Corporation, MJM&M Oil Company and Standard Oil Company of California to drill a test five miles north of the San Emidio Nose field and six miles northeast of the Louden property. The hole was dry at 13,600 feet.

Correspondents hinted at a big play being put together to the north of the Louden property. Louden eagerly seized on this as the play that would turn interest to his area and finally bring drilling to his lease. The play seemed to take an agonizingly long time to put together. In April, 1961, Ferguson & Bosworth drilled the Shell Fee No. 1 on Sec. 30, 32S-26E, Kern County, two and one-half miles north of the Louden property. Others with interest in the wildcat were Norris Oil Company, the discoverer of Cuyama Valley's first oil field, and Pacific Oil & Gas Development Corporation, San Francisco. The hole was dry at 13,617 feet.

While wildcatters were running into trouble, Richfield at San Emidio Nose was having its problems. After a string of some 20 completions, the company moved in a Coastal Drilling Company rig to drill an extension test one mile northwest of the field and came away with a duster at 13,960 feet. The hole was abandoned in March, 1961. Before the year was out, the company had abandoned three additional extension tests, and the San Emidio Nose field was regarded as fully defined.

On May 26, 1962, four years after Richfield Oil Corporation had leased the Louden land, with one year left to go on the five-year lease, a certified letter arrived at the Louden home in Fairfield. The letter read:

"Dear Lessor:

Richfield Oil Corporation, as Lessee under that certain Oil and Gas Lease referred to in the enclosed duplicate original quitclaim deed, has elected to exercise its right to surrender and quitclaim said lease. The enclosed document, which gives effect to such quitclaim, is furnished for your file.

We wish to take this opportunity to thank you for your past cooperation in all matters pertaining to this lease."

The letter was signed by J.F. Reynolds, land and lease.

In the May 31, 1962, issue, *The Bakersfield Californian* noted the quitclaim and Louden's long campaign to get a well drilled, stating:

"It was a rare Kern County oil company, petroleum engineer, consulting geologist, drilling contractor or service company representative who had not heard at least once from the Iowan.

"Time ran out this week as Louden, together with many other landowners in the area, received quitclaim notices from Richfield advising that the oil company was giving up their land.

"The company had prospected without success between the San Emidio Nose field's 25 wells (5,850 barrels-per-day oil production) and the still-unproved area to the northwest. The decision obviously was that further drilling could not be justified at this time.

Oil joined cotton as a contributor to the rich yield at Yowlumne. Loffland Brothers Company's Rig No. 92, background, drilled the well that found oil beneath the John Louden parcel. (William Rintoul)

"Will another company take up the exploratory burden?

"Back in Fairfield, Iowa, John P. Louden, still a booster for Kern County oil, still convinced there's oil under his land, is waiting, hopeful that the next company to lease his land will move a rig into the field and find the elusive oil he's sure is there."

On the same day that he received the notice Richfield was giving up its lease, Louden sat down at his typewriter and wrote letters to Bill Croghan, Standard Oil Company of California; Herb Harry, Union Oil Company of California; and Alex Sarad, Tidewater Oil Company. The three headed land offices of their respective companies, and to each, Louden offered a lease on his land. The three responded courteously. The answer was the same in each case. None of the companies was interested in leasing the Louden property at the present time. Louden submitted the parcel to Mobil for consideration. He owned 120 shares of the company's common stock. Mobil's negative response struck Louden as cavalier. He sold the stock.

On July 23, 1962, Louden wrote to a friend in California: "I know they have all turned me down on drilling in our area but they will be back and glad to drill in due time."

Though the bloom was gone, Louden continued to write with, if anything, more determination than ever. In California, interest shifted away from San Emidio Nose. Some who had been reasonably good in answering his letters found themselves writing less and less often, if at all. There simply seemed little or nothing to say.

Another element entered into the one-sided correspondence, and it too, unfortunately, had a tendency to diminish further the volume of mail traveling from California to Iowa. At one point during the active development of the San Emidio Nose field, the *Daily Midway Driller,* to which Louden had subscribed as a means of helping to keep up with developments, failed to appear for a period of almost a month in Louden's post-office box, which, as luck would have it, was directly over the box held by Louden Machinery Company. After puzzling over the paper's absence and checking to be sure he had paid his subscription, Louden inquired at the post office and discovered that the newspaper had been placed by mistake in the plant's post-office box. The incident planted the seeds of doubt in his mind. Was his mail going astray? Was that the reason he did not hear from correspondents in California?

Louden worried. He added a line to his address, just below his name, writing "Not at the Plant," so that when people like R.G. Follis, who was chairman of the board of directors of Standard Oil Company of California, responded to one of his letter, the oil executive's secretary carefully addressed the heading at the top of the letter to Mr. John P. Louden, Not at the Plant.

To circumvent what he decided was a massive failure in the mail system, Louden began sending letters special delivery airmail, sometimes by certified mail, and this, too, antagonized some when they were awakened on a Saturday or Sunday morning by the mailman at the door, asking them to sign for a letter that wanted to know what was happening, particularly since nothing was happening. It also led to a situation which, for the uninitiated, could be misleading. In writing to various oilmen, Louden sometimes would remark, perhaps as a means of spurring interest, that on that very day he had received a special delivery letter from a prominent oil executive, which tended to give the impression some vital business was being transacted, without recognizing that the priority mail had been enclosed for the answer by Louden himself. In another move to ensure the receipt of mail, Louden rented postoffice boxes in such nearby towns as Battavia and Mount Pleasant, Iowa, hoping thereby to escape what he began to suspect was deliberate interference with his mail.

In the effort to enlist support for the drilling of a well on his property, Louden proved resourceful. In a particular issue of *The Kemper News,* a tabloid published biweekly by the Cadets of the Kemper Military School and College, Boonville, Missouri, Louden read that Charles R. Walbert had been named to the Kemper Board of Trustees. Walbert was identified as an oil producer and president of Charles R. Walbert & Company, Oklahoma City, Oklahoma. Louden remembered him as a classmate when he attended Kemper in 1927-1929. Walbert had been a drummer at the same time that Louden was a bugler in the Kemper bugle corps. Louden lost no time writing Walbert, advising him of the situation with regard to the land in California and asking Walbert's help. Walbert responded, saying he would contact a friend in California to see what, if anything, could be done. The friend, Walbert said, was watching some wells near Sacramento, but as soon as he got back to Los Angeles, he would check the area and get all the dope together. Walbert spoke of putting together a block to get a well drilled.

Another Kemperite, and chairman of the board of trustees, was Louden's classmate Charles J. Hitch, who was assistant secretary of defense, Washington, D.C. Louden wrote, asking if there was anything the government could do as a matter of ensuring a supply of oil for national defense. *The Kemper News* also mentioned Arthur B. Ramsey of San Francisco, describing him as a California oilman. Louden thought he remembered Ramsey from the Kemper bugle corps. It turned out it was Ramsey's brother Charley whom Louden had known.

Among other affiliations, Louden was a member of the Shrine. At a meeting of Kaaba Temple in Davenport, Iowa, he met Representative Fred Schwengel, home from Washington, and this led to an exchange of correspondence over the prospect of getting a well drilled on the California land. As a Shriner, Louden supported the Shrine's work with crippled children. He decided if someone did bring in a well on his property, he would donate a one-third share of the royalty to the Shriners' hospital fund. He communicated this offer to a fellow Shriner in Los Angeles, District Attorney Evelle J. Younger, and Younger responded, lauding him for the offer.

Not long after Richfield had quitclaimed the lease, Louden, on a trip to Rochester, Minnesota, for a routine checkup at the Mayo Clinic, had a chance meeting in the lobby of the Kahler Hotel, where both were staying, with Randolph Scott, a well-known Hollywood actor who had starred in such movies as *The Last of the Mohicans, Virginia City, When the Daltons Rode, Western Union,* and *Gunfighters,* among others. Scott, too, was in Rochester for a checkup, explaining to Louden, "A stitch in time saves nine." In the course of a half-hour visit with the actor, Louden learned that Scott participated in oil ventures and told him of his property. Scott said he would have his friend Ken Evans, a consulting petroleum engineer in Bakersfield, look into the matter.

On his return to California, Scott wrote Louden:

"I have just looked over Mungers Map in locating your 80 acres and found that you are surrounded to the northwest, north and northeast by dry holes. These are deep tests and expensive. To the south in the San Emidio Nose field, these productive wells are likewise deep tests, ranging from 12,000 feet plus to 14,000 feet plus. The fact that I am referring to these wells is important in the sense that just to pick up an 80 acre tract for a deep test is certainly anything but an economic move. I can understand how you have had difficulty in leasing this 80 acres.

"Anyway, I expect to see Mr. Evans within the next week and will ask him to give you a brief engineering opinion on the possible productiveness of your 80 acres.

"I am glad you and your wife passed the tests at the Clinic. Likewise, I was fortunate enough to come through OK myself. In the meantime, good luck."

Weeks stretched into months and months into years, and the list of companies and individual operators that had turned down Louden's offer to lease his land grew until it included virtually a who's who of California producers. One day it was Humble Oil & Refining Company politely declining, another Texaco, another Phillips Petroleum Company, another Buttes Gas & Oil Company, another Pan Petroleum Company of Denver. The list of correspondents declined until only a handful still wrote, among them Roy Loudon, Jim Radoumis, Mike Rector and Gene Winer in Bakersfield and Phil McCormick in Taft.

The rejection by oil companies was not the only blow. *Oil & Gas Journal,* which periodically reviewed its subscription list, seeking to weed out those who were not directly connected with the oil industry, refused to renew Louden's subscription at the usual rate, insisting that he pay $25 a year, a nonindustry rate, instead of the $5 a year that those in the oil industry paid. Louden insisted, to no avail, that his ownership of stock in oil companies and of potential oil land in California should qualify him for the industry rate. The magazine stood fast, and Louden quit his subscription, wondering in letters to friends how it was that college students could subscribe for $3 a year while he was supposed to pay $25 a year.

Mail, when Louden received any, was generally negative. In March, 1965, a representative of Standard Oil Company of California wrote, "We have previously reviewed the geology in the area of your property and must again confirm that we cannot establish any exploratory interest in drilling on or around your properties."

The well that proved John Louden was right—Tenneco Oil Company's Louden No. 47X-34, a joint interest well with Texaco. The well was the first of two fast-flowing wells drilled by Tenneco to develop the 80-acre Louden parcel. (William Rintoul)

In one burst of letter-writing activity, Louden wrote 25 letters, sending stamped, self-addressed envelopes with each letter. He did not receive a single reply.

Though the situation seemed hopeless, Louden contined to believe in his property. Occasionally, there would be a glimmer of hope. Someone would write something that indicated all of the geologists had not given up.

Archer H. Warne, senior geologist with Richfield, wrote: "I think interest will swing back again to deeper testing on the San Emigdio Nose structure where your property is located. I'm back on old projects again, including the Maricopa Flats region, and will surely keep your property in mind as the work advances. We have several new men in the office now, and I believe in the idea that new blood generates new ideas. Well, it's back to the drawing boards again and hope to continue to hear from you."

Tom Roy wrote from Casper, Wyoming, where he had been transferred by The Ohio Oil Company: "It would certainly seem there will be some more stratigraphically-entrapped oil found along the east-west extent of San Emidio Nose, and the north side, on which you're located, should have the best chance. However, it will take some-body who's working the area and who has the imaginative skill to devise a trap to do the job."

In 1967, Tenneco Oil Company took over Kern County Land Company and overnight became a new presence to be reckoned with in the California oil picture.

Louden quickly wrote to B.D. Carey, Tenneco's chief geologist in Houston. Carey replied, saying that Tenneco meant to take the cover off the book in the area where Louden's land was situated as soon as the company could get rolling in Bakersfield. As a measure of Tenneco's inter-est in California, the company assigned O.W. "Tiny" Ward, a top production man, as vice president and general man-ager to ramrod the Bakersfield operation.

Louden wrote Carey advising him not to listen to "die-hards," for if the company relied on such advice, Louden said, instead of its own judgment, it would never drill. Of the doubters, Louden wrote, "They do not know, and they do not want anyone to find out if they are wrong."

From Houston came word to be patient, that Tenneco meant to check out the area by looking at it with the company's own way of thinking, not what was cut and dried. Louden enthusiastically bought 75 shares of Ten-neco stock. He wrote to a friend in California, "I think the company is going to have a great future."

On Christmas Day, 1968, Louden with his wife, Helen, flew from Fairfield, Iowa, to Rochester, Minnesota, where Louden was to enter the Mayo Clinic. He had not been feeling well, suffering stomach pains and general debility he described in a letter as "training on what happens to a human as they get older."

On January 7, 1969, John P. Louden died. The cause of death was diagnosed as malignancy of the liver.

By 1972, three years after Louden's death, Tenneco Oil Company under the direction of Don Taylor, California exploration manager, had put together a deep play. The Louden property was leased from Helen Louden as part of the block. Tenneco turned the play to Texaco Inc., which moved in two and one-half miles southeast of the Louden parcel to drill what was programmed to be a 25,000-foot test, using Loffland Brothers Company's Rig No. 34. There was a fishing job at 20,704 feet. The operator stopped at that depth and began testing back up the hole. In January, 1974, Texaco announced that the San Emidio No. 1 on a 24-hour gauge had flowed 428 barrels of 33-gravity oil and 500,000 to 600,000 cubic feet per day of gas through a 26/64-inch bean. The production was from Upper Miocene Stevens sands between 11,305 and 11,465 feet.

The discoverers of the new field turned to Richard C. Bailey, director of the Kern County Museum, for a name. Bailey suggested Yowlumne after a tribe of Indians he had described in his book *Heritage of Kern,* published in 1957 by the Kern County Historical Society. The name, meaning "Wolf People," was that by which the inhabitants of the site of what later became Bakersfield had been known among their fellow Yokuts.

Following the discovery, Tenneco took over as operator, moving the productive limits of the Yowlumne field ever closer to the Louden property.

On January 13, 1979, Tenneco spudded in with Loffland's Rig No. 92 to drill the Louden No. 47X-34 as the first hole on the property. Drilling went smoothly. Thirty-nine days after it had been spudded in, the well was at total depth of 12,943 feet. En route, the drilling bit had encountered some 300 feet of oil sand in selected intervals in the Stevens section. One geologist described the pay zone as the richest yet encountered in the Yowlumne field. In March, Tenneco completed the Louden well flowing 1,008 barrels a day of clean, high-gravity oil.

12 The Day the World Stopped

The Bluewater No. 2, drilling for Humble Oil & Refining Company on a sparkling sea, helped discover major oil reserves in the Five-Mile Trend some 20 miles west of Santa Barbara. (William Rintoul)

In the predawn darkness of Saturday morning, September 14, 1968, the supply vessel *Oil City* put out from Stearns Wharf in Santa Barbara following a heading south-by-southeast into the Santa Barbara Channel. The vessel seemed somewhat top-heavy, which gave it a tendency to roll, but the channel waters were smooth and the voyage was without undue roughness. There was the smell of seaweed and salt water in the cool air and the sky, though dark, appeared clear, promising another of those beautiful end-of-summer days for which Santa Barbara is famous, the kind of days when the town becomes a weekend retreat for those seeking a quiet, sun-filled rest away from it all, without the foggy mornings that are common in the summer when most vacationers are there.

Aboard the *Oil City* were some twenty or thirty men who had gotten up sleepily in the small hours of the morning in their homes in such cities as Santa Maria and Los Angeles, Bakersfield and Pasadena, to drive along mostly empty highways to arrive in time for the scheduled 5:30 A.M. departure of the supply vessel, which had come up to Santa Barbara the night before from its home port of Port Hueneme, 35 miles to the southeast. The men were mostly oil engineers and most were associated with Union Oil Company of California. They were informally dressed, enjoying the outing, for they were there as observers rather than participants. The host aboard the vessel was Bill Huskey, who saw to it that there was plenty of hot coffee and Danish rolls for those who wanted them. There were also a galley and cook for those who wanted a more substantial breakfast. Huskey was general manager of Port Hueneme Industrial Service Inc., which had been formed by Howard H. Bell Jr., president of Bell-Western Oil Company, a Bakersfield-based independent producer. The service company was developing a 27-acre spread at Port Hueneme as an offshore service center to serve the boom that seemed to be coming off California.

The *Oil City* lumbered along through the smooth water, rolling lightly. The vessel was not fast, but it was sturdy and one had the feeling it could survive rough seas. About forty-five minutes out into the channel those on deck saw looming ahead in the grayness of the dawn the outline of what appeared to be the frame for a huge structure, lying on its side on a long low barge. The vessel approached, and one could see the tall crane of a derrick barge and several workboats standing by. The enormous structure dwarfed everything nearby. As it rode serenely on the barge, harsh spotlights pierced the gloom to fashion a crazy quilt of light and shadow over the structure. If it had not been so large, one might have taken it for a fabrication from a children's erector set.

The 1,700-ton structure, newly arrived after a six-day trip by barge out of Portland, Oregon, was the jacket for Union Oil Company of California's Platform A, which would be the first drilling and production platform to be positioned in the Santa Barbara Channel following the sale of Outer Continental Shelf leases by the federal government in February. The structure was the height of a 15-story building. It had been built in Vancouver, Washington, by American Pipe & Construction Company, and was to be the first of three offshore platforms that Union planned to position on an 8.2-square-mile tract some six miles offshore from Santa Barbara. The jacket comprised the legs and substructure for Platform A, which was designed to accommodate 56 wells.

After the *Oil City* had steamed close enough to give those on board a good look, the vessel moved away, standing by for further developments, which were not long in coming.

The sun, blocked by misty fog, was still not visible as the crew aboard the barge began to winch the jacket into the water. The jacket slowly tilted up from the barge, poised momentarily and then, at 7:01 A.M., slid quietly into the mirror smooth water, 190 feet deep at this point. There was no big splash. The jacket simply slid into the sea.

Aboard the *Oil City* C.H. Chadband, who headed the Union Oil Company of California offshore task force for

258

the channel venture, watched nervously, leaning so far over the rail that one co-worker said he looked as if he was going to jump right in after the platform.

The structure slowly bobbed to the surface, riding low in the water and on its side, as it had been designed to do, kept afloat by air sealed in its 12 steel legs and by two steel flotation cells. The 250-ton capacity derrick barge *H.A. Lindsey*, back in California waters after a 42-day tow from the Gulf Coast, moved in, and a crew attached a cable to begin the careful job of righting the jacket.

After the jacket had been righted, the derrick barge held a strain on it, keeping the jacket legs off bottom, while the tug *Apollo* out of San Francisco tied on to one leg and rotated the jacket into the desired position in the water. To assure an accurate fix, there followed several hours of surveying and trial positioning, with surveyors using line-of-sight, shoran and laser beams.

While engineers on the *Oil City* speculated on why there had been no big splash like there had been four years earlier when Platform Eva had been launched off Huntington Beach, the cook quietly slipped out of the galley with a

The jacket of Union Oil Company of California's Platform A slid quietly into the Santa Barbara Channel, creating scarcely a splash in the calm water. (William Rintoul)

fishing pole and tried his luck off the side of the boat. Having no success near bottom, he took the sinker off his line and tried casting out and reeling in, hoping to attract fish with a surface lure. He did not catch any fish.

By noon, the jacket had been lowered into place. It was anticipated that the derrick barge, operated by Mid-Valley Inc., would spend some six weeks preparing the platform for its role. Brown & Root, the contracting firm in charge of the turnkey project, would have to drill pilings into the ocean floor to secure the jacket in place. Then production and drilling decks would be lifted on top to make room for several thousand tons of equipment, including such things as separators, cranes, drill pipe, mud tanks, motors and crew quarters. Meanwhile, other contractors would be laying pipelines to shore and a 34,000-volt electric cable to power the drilling and production equipment.

The platform would mount two drilling rigs, including a conventional rig and a rig tilted to a 30-degree angle. The tilted rig would be necessary to broaden the drilling reach in developing extremely shallow oil sand encountered beneath the tract.

Union had announced the first discovery in March, only six weeks after the sale. The company said a test well near the center of the tract had flowed 27.8-gravity oil at a rate of 1,800 barrels per day from Repetto sands between 2,000 and 2,700 feet. The four-hour test had been through a half-inch choke with a flowing tubing pressure of 269 pounds per square inch. The company had drilled the hole from Western Offshore Drilling & Exploration Company's *WODECO I,* a drilling barge, in 190 feet of water. The discovery had been confirmed by a second well drilled from Sun Marine Drilling Corporation's *George F. Ferris,* a jackup (a drilling vessel on which the drilling platform is attached to adjustable legs which rest on the seafloor). Drilling had confirmed a long but shallow trend of oil sands that were believed to hold more than 100 million barrels.

On the *Oil City,* after the jacket had been positioned, everyone suddenly seemed hungry. The cook, fortunately, had not relied on success with a fishing pole. He had prepared a standing rib roast, plus apple pie and all the trimmings necessary to satisfy appetites whetted by salt air and watching a job well done. As the vessel pulled away from the platform for the return to Santa Barbara, those on board fell to with pleasure, enjoying a wonderful meal. The day had turned out to be another beautiful day, living up to its early promise, with a clear blue sky and gentle warmth, a day to relax and enjoy life. There was a sense of well being about the whole affair as, in a way, there was about the oil business itself in California.

If things were going well in the Santa Barbara Channel, they were going equally well in the California oil industry. The industry was in the process of establishing an all-time high for California production. From the dark days that had followed the setting of the existing record in 1953, the industry had suffered a painful seven years in which each successive year's production had dropped below the year before. Part of the decline could be blamed on the shutting in of wells when California oil could not compete with cheap oil from overseas, but a major share of the blame had to lie with the fact that California's established fields were declining and wildcatters were not finding new production fast enough to replace the decline.

There had been a turnaround in 1963 when the massive waterflood initiated at Wilmington to halt land subsidence and also to increase oil recovery had begun to show big gains. It was conservatively estimated that the field that year was producing some 40,000 barrels a day over what it might have produced without waterflooding, or almost 100 percent above the normal primary production rate.

Though there had been a slight drop in production in 1964—to 300 million barrels instead of the 300.8 million barrels of the year before—there had been a production increase each year since until now, in 1968, producers were on their way toward setting an all-time high record. When the final production figures would be added up at the end of the year, they would show that operators had set a record by producing 373.2 million barrels, or approximately 1,025,000 barrels daily, topping the 367.3 million barrels, or approximately 1,009,000 barrels daily, that had been produced in 1953.

One out of five barrels of the oil that was propelling California to a production peak in 1968 was coming from offshore, or more than 210,000 barrels daily. The biggest single helping was from the THUMS Long Beach Company development of the seaward extension of the Wilmington field. The company was past the midway point in a projected 800-well drilling program that had been started in the summer of 1965. THUMS had developed a production of 130,000 barrels daily and was running 16 contract drilling rigs on the four Astronaut Islands, pushing production on toward an anticipated peak of more than 170,000 barrels daily.

More offshore oil was coming from fields off Southern California, totalling about 52,000 barrels a day. The fields included Belmont Offshore, Huntington Beach, Newport Beach, Torrance, Venice Beach and West Montalvo. The biggest share came from the South area offshore of the Huntington Beach field, which was contributing some 37,000 barrels daily.

Additional offshore oil was coming from fields discovered in the Santa Barbara Channel. The fields were contributing some 30,000 barrels daily. They included Alegria Offshore, Carpinteria Offshore, Conception Offshore, Cuarta Offshore, South Elwood Offshore and Summerland Offshore. All had been developed on state-owned tidelands with exception of a portion of Carpinteria Offshore field, which was under development from a federal tract opened for leasing when it became apparent the federal acreage would be drained by wells on adjoining state-owned land.

Steam was another significant factor in the resurgence of California production, adding an estimated 130,000 barrels daily to the state's output. More than 400 steam generators were working, principally in fields that produced heavy crude from shallow depths. Three out of four generators were concentrated in three fields: Kern River and Midway-Sunset in the San Joaquin Valley and San Ardo in

Union Oil Company of California's Platform A, escaping oil and gas bubble at right of platform and Santa Barbara County beaches in the background—these formed the elements of a bitter controversy that would have a profound effect on the development of offshore oil resources. (William Rintoul)

the Salinas Valley. Thanks to steam, production showed substantial gains in each of the fields. In the six years since steam had begun to gain a foothold, production at Kern River had climbed from 24,700 barrels daily in 1962 to 68,500 barrels daily in 1968, a gain of 43,800 barrels per day. At Midway-Sunset, production had climbed from 41,600 barrels daily to 92,300 barrels daily, an increase of 50,700 barrels per day. At San Ardo, the rise had been from 30,500 barrels daily to 36,700 barrels daily, a gain of 6,200 barrels per day.

Though steam promised more increases, offshore held the spotlight, especially in the Santa Barbara Channel, where the federal government in February, 1968, had opened the Outer Continental Shelf to oil exploration.

There had been drama in the lease sale on Tuesday morning, February 6, in the Renaissance Room of Los Angeles' Biltmore Hotel. The sale had been preceded by a long and bitter battle between the state of California and the federal government over who owned the channel waters. The struggle had been resolved in the federal government's favor. There had been strong opposition, too, from those who did not want drilling under any circumstances in the channel, but they too had not been able to stop the sale.

In the Renaissance Room, sealed bids were tendered at 9 A.M. and publicly opened one hour later. Even before the first bid was opened, one could feel tension building as some 500 bidders and other interested persons took their places. As the deadline neared, conversation rose to a pitch that might have competed with the Chicago grain market.

There was speculation that as much as two billion barrels of oil, maybe more, might lie beneath the 110 tracts that were being offered. Some tracts were under as much as 1,000 feet of water. Part of the offshore land lay on the Rincon trend just south of the Summerland Offshore field. On virtually every company's geological map, two tracts within this area hinted at major reserves of oil and spectacular bidding.

Underscoring tension in the room was the ominous knowledge that the preceding year's Mideast crisis had severely disrupted traditional lines of crude supply, increasing the pressure for assured domestic supplies. There were national security considerations, too, including the Vietnam War, the Pueblo incident and the nagging dollar drain that was prompting Americans to spend their money at home.

The rules of the sale were simple. Those interested in leasing offshore tracts were asked to submit sealed bids offering bonuses for the tracts they wanted. There would be a rental price of $3 an acre annually, plus 16.67 percent royalty on any oil produced.

At precisely 10 A.M., a hush fell over the audience as William Grant, Los Angeles office manager for the Interior Department's Bureau of Land Management, opened the sale. Speaking slowly and clearly, Grant began reading off bids. Bidding was relatively modest for the first tracts. Inasmuch as the higher-numbered tracts were in the best-looking areas, the price of bids increased as the meeting progressed.

Humble Oil & Refining Company proved to be the front runner, ultimately paying out almost $218 million to win interest in a total of 34 tracts, including those in the greatest water depths.

Twenty-five companies competed for leases, offering bonuses which at times made it appear as if they wanted to see who could have the privilege of paying off the national debt.

Suddenly there was an expectant hush, indicating that the two tracts everyone had been waiting for were coming up. A joint venture headed by Superior Oil Company and Marathon Oil Company took the first of the two tracts for a bid of $38 million.

There remained only Tract 402. The first bid opened came from a bidding group that included Continental Oil Company, Cities Service Company, Pan Petroleum Company Inc. and Phillips Petroleum Company. It was for $18 million. Humble Oil & Refining Company came next. Bidding alone, the company raised the price to a whopping $55 million. By this time, Shell Oil Company's bid of $22 million was not competitive.

There was only one bid left. Land Office Manager Grant opened the bid from Gulf Oil Corporation, Texaco Inc.,

Stockpiled material for Platform A was stored in a huge new warehouse built by Port Hueneme Industrial Service Inc. at Port Hueneme. The large bits, center, were to be used to drill in the platform legs. (William Rintoul)

Mobil Oil Corporation and Union Oil Company of California. All eyes focused on the representatives from the four companies as the figure was read off: $61,418,000, or $11,373.70 per acre. The bid was the highest of the sale and the biggest single bid ever offered for an offshore lease, nearly double the previous record of $32,500,000.

For a moment, the audience sat silently, as if stunned. Then wild cheering broke out as winners relaxed from the tension that had been building up over the past weeks. In a 90-minute-long sale, oil companies had offered the federal government $603 million for the right to drill in the Santa Barbara Channel. It was the greatest total of high bids ever received in a lease sale.

Union Oil Company of California had been a participant in winning bids for eight tracts. When the company's Pacific Coast Manager John R. Fraser, Manager of Exploration George Pichel, Manager of Lands Herb Harry and Corporate Assistant Treasurer Bill Craig added up Union's share of the total high bonuses, they made an interesting discovery. Union was known for its Seventy-Six trademark; the company's share of the total high bonuses came to $76 million. It appeared to be a good omen.

On Platform A, the first well was spudded in mid-November, the second a short time later as Union initiated active development of what had been named the Dos Cuadras field. The name, in Spanish, meant two blocks, taking note of the fact that the field lay beneath the block held by Union and partners and, in smaller measure, the adjoining block on the east held by Superior and its partners. The drilling contractor for the two-rig operation from Platform A was Peter Bawden Drilling Company Inc.

In that same month, Union positioned the jacket for Platform B in 192 feet of water one-half mile west of Platform A and made application to the Army's Corps of Engineers for a permit to position Platform C one-half mile farther west.

The start-up of drilling from Platform A and the preparations to expand the operation to two more platforms were overshadowed by tragedy. In early morning hours on

Monday, November 25, the *Triple Crown*, 170-foot vessel serving as an anchor handling and supply vessel for Deep Water Operators Inc., sank in 270 feet of water while moving anchors for Santa Fe International Corporation's Bluewater No. 2, a semisubmersible which was drilling for Humble Oil & Refining Company eight miles southeast of Santa Barbara. The crew aboard the vessel had picked up all eight anchors and was reeling in the chain of the last anchor when the stern settled. The boat capsized to starboard and sank. Sixteen aboard were rescued. Nine men lost their lives. The steel-hulled ship, on the job just three weeks after being built by Burton Shipyard in Port Arthur, Texas, had been described as the world's most powerful combination tug-supply boat. It had been built especially to lift and lower the anchors that kept floating drilling vessels in position over drill sites.

By mid-January, 1969, three wells had been completed from Platform A and shut in, a fourth was being drilled and a fifth had been started. The fifth well was No. 402-A-21. The rig was the vertical rig.

Crews drilled 12¼-inch hole to a depth of 244 feet, opened the hole to 17½-inches in diameter and ran 13⅜-inch surface casing to a depth of 238 feet, cementing the pipe solidly from the shoe to the ocean floor.

Blowout prevention equipment was installed, consisting of a double Shaffer gate with blind and 4½-inch pipe rams and a G.K. Hydril. The equipment offered the capability of shutting in the well regardless of whether there was pipe in the hole. The equipment was tested before the crew

Up to 1,000 men worked to clean Santa Barbara's beaches of the crude oil that washed ashore following the blowout at Platform A. (William Rintoul)

drilled out below the shoe of the casing. Crews cut 12¼-inch hole directionally to a total measured depth of 3,203 feet below the ocean floor.

On Tuesday morning, January 28, the drilling crew started to pull the drill pipe from the hole in order to log the well. The first five stands pulled with difficulty, but the string pulled free and the next three stands were removed easily. At 10:45 A.M., while the crew was disconnecting the eighth stand, gas started blowing through the drill pipe.

In the tense minutes that followed, the crew attempted to install an inside blowout preventer at the top of the drill pipe. The well was blowing too hard. The effort failed. The crew started to pick up the kelly to stab it into the drill pipe. The rotary hose caught on a standpipe fitting and broke it off. The fire hazard was considered too great to pursue attaching the kelly. The crew dropped the drill pipe down the hole so that the blind rams could be closed to shut in the well. The rams were closed at 11 A.M., 15 minutes after the blow had begun. Normally, this would have taken care of the situation.

Soon after, oil and gas began to bubble to the ocean surface at no less than four different spots. The biggest

From Stearns Wharf in Santa Barbara, workmen loaded portable tanks to hold the heavy drilling mud that would be required to control the blowout at Platform A. (William Rintoul)

bubble was by the platform base. Oil and gas, stopped from coming up through the well, appeared to be coming from a fracture line on the ocean floor.

The crew made an attempt to stop the flow by pumping mud that weighed 90 pounds per cubic foot down the casing and bleeding off the gas. The attempt was unsuccessful.

The crew mudded in the three wells which had been completed from the platform.

Escaping gas delayed the effort to reestablish circulation in the wild well in order to kill the blowout. Crews had to wait for favorable winds to disperse the gas.

An additional G.K. Hydril preventer was installed. The crew began to run drill pipe into the hole. An inside check valve, installed at the bottom of the drill pipe, prevented upward flow of oil and gas.

By late the next day, crews had succeeded in connecting drill pipe to the pipe that had been dropped into the hole the previous day. An attempt was made to circulate mud into the hole through the drill pipe. The attempt failed, even at a pressure of 5,000 pounds per square inch. The crew tried to pull the drill pipe from the hole. The pipe moved about seven feet but did not pull free.

While oil crews battled the blowout at Platform A, skirmish lines formed onshore for battles that subsequently would prove far more complex than that faced in subduing the well that had gotten away.

A slogan provided a rallying cry and a name for an organization that sprang into existence in Santa Barbara. The slogan was "Get Oil Out," the group's name, GOO. Members circulated petitions calling for an end to oil operations in the channel. Among the early signers were Senator Edmund Muskie (Democrat-Maine) and California Assembly Democrat Leader Jess Unruh (Democrat-Inglewood), who inspected the oil slick spreading from Platform A from a Coast Guard plane.

James Bottoms, founder of GOO, announced that the following Monday, February 3, would be "Black Monday" and urged motorists to boycott oil firms involved in the offshore drilling activity.

Dr. Edgar Wayburn, president of the 69,000-member Sierra Club, headquartered in San Francisco, wired President Nixon and Interior Secretary Hickel demanding an immediate halt to all drilling in the channel.

Representative Charles M. Teague (Republican-California) appealed for a permanent halt to drilling in the channel "no matter what the monetary cost to the federal government may be."

Ellen Stern Harris, representative for the public at large, Los Angeles Regional Water Quality Control Board, asked the California Water Resources Board for a resolution asking the Army's Corps of Engineers to halt all offshore drilling operations, both on federal and state tidelands.

Gerald Firestone, mayor of Santa Barbara, said of his constituents, "You bet they're mad. This city is known throughout the world as an attractive community. Look at it now."

On its editorial page, the *Los Angeles Times*, the most widely circulated newspaper in California, stated:

"The fundamental question raised by the current crisis is one of the values and priorities.

"Is the nation so in need of new oil supplies or the U.S. Treasury so anxious for revenue that California's priceless shoreline should be imperiled?"

On the nation's prime time television news programs and on the front pages of the nation's newspapers there appeared pictures of dead and dying sea birds. The Associated Press reported that scores of dead birds—gulls, godwits, curlews, black-bellied plover and grebes—were

Looking like an invasion fleet, barges and work boats assembled for the assault. (William Rintoul)

found along a 14-mile beach front from Goleta to Rincon Point.

There was an angry demand for an end to offshore drilling.

On Platform A, men worked to control the blowout in an atmosphere that posed a continuing threat of explosion and fire, mindful that there had been no outcry for shutting down channel operations two months before when the *Triple Crown* went down, taking nine men's lives.

To control the blowout, Union Oil Company of California mounted a two-front effort. One campaign was waged from Platform A, where engineers hoped to bring the well under control with mud. Another campaign was mounted from the *WODECO II*, a drilling barge hastily summoned to the scene from Long Beach.

From *WODECO II*, positioned 1,000 feet south of the platform, a drilling crew spudded in on Sunday morning, February 2, five days after the blowout had begun, to drill a relief well for use if control efforts from the platform failed. After cementing 20-inch surface casing at 592 feet, the crew drilled ahead to intercept the well that was blowing out at a depth of about 3,000 feet.

On the platform, crews after failing to pull the drill pipe free made several attempts to unscrew the pipe at a point deep enough to circulate mud into the hole to kill the flow. Each time the pipe disconnected at too shallow a depth.

Crews milled out the inside check valve and ran a perforating gun down the string of drill pipe to a depth of 2,666 feet beneath the ocean floor. The pipe was perforated with 97 holes in the interval from 2,584 to 2,607 feet.

While equipment was being rigged to pump mud down the drill pipe, sea water was pumped down and out the perforations at a rate of 23 barrels per minute at a presssure of 2,100 pounds per square inch. The sea water failed to curb the flow of oil and gas.

On Tuesday, February 4, a storm hit the besieged Santa Barbara area with 35 miles-per-hour winds and 15-foot seas, delaying the final assault on the well for more than a day. The storm broke up the oil slick formed by the blowout, scattering oil on Santa Barbara County beaches, with other debris washed into the surf by onshore floods.

The outcry to halt offshore drilling intensified.

On Wednesday, February 5, crews on Platform A pumped 250 barrels of 100-pound mud down the drill pipe. That night they pumped down sea water, continuing through part of the next day while additional pumping units were being set up and the operator was waiting for a barge with an additional supply of mud.

After the newly installed pumping units were tested, 1,000 barrels of sea water was pumped down the drill pipe, followed by 210 barrels of 116-pound mud pumped at a rate of 27 barrels per minute at a pressure of 2,650 pounds per square inch. A line blew off the wellhead. It was necessary to shut down the pumps.

After repairs were completed, the mud in the drill pipe was displaced with sea water and the well was shut in. During the entire pumping operation, the casing pressure of the well remained at 190 pounds per square inch. However, after the mud had been pumped into the hole, the large bubbling area at the east end of the platform grew noticeably smaller. After 10 minutes, it returned to its original size.

Meanwhile, Union had lined up a D-Day fleet of barges and workboats, portable tanks and other necessary equipment, putting together a mud deliverability of approximately 15,000 barrels.

When weather cleared on Friday, February 7, ten days after the blowout began, the fleet converged on Platform A. Steady pumping began at 4 P.M. By 5 P.M. nine units were in operation pumping mud down the drill pipe at a rate of 30 barrels per minute—or 1,800 barrels per hour—at a pressure of 3,750 pounds per square inch. At this time, mud was pumped down the annulus outside the drill pipe, utilizing the rig pumps. By 5:30 P.M. the gas bubbles began to decrease in size. The well was killed after 13,000 barrels of 90- to 110-pound mud had been pumped down the hole.

The coup de grace was administered with 900 sacks of cement, plugging the well for abandonment.

Bringing the blowout under control, of course, was only one part of the problem. During the uncontrolled flow, it was estimated by the operator that oil was escaping at a 500-barrels-per-day rate. There was, however, no way to gauge effectively the huge bubbles, and estimates of the flow ran higher and lower, often depending on the estimator's affiliations. Some industry observers with experience in production from other fields in the state-owned portion of the channel estimated the flow of oil at upwards to 2,000 barrels daily.

The oil formed a slick that, like the rate of uncontrolled flow, was estimated at widely varying figures, ranging from 200 to 800 square miles. The effort to deal with the slick began within hours of the time the blowout occurred.

Humble Oil & Refining Company, drilling the Five-Mile Trend with the WODECO IV, proved up major oil reserves in the Santa Barbara Channel. (William Rintoul)

The campaign to eliminate the slick took several forms and in its initial stages underscored the fact that industry had not developed adequate tools to handle large oil spills. It proved an ironic turn to the Outer Continental Shelf oil search that the first oil ashore in the wake of the year-old lease sale that saw the federal government reap almost $603 million in bonuses should come ashore on the tide, despoiling some of the highest-priced beach front acreage in California.

One of the first phases of the attack on the floating oil involved putting a flight of three crop duster-type planes into the air to spray dispersant over the oil. Results, according to various observers, were minimal, and the effort largely was curtailed following conservationists' complaints that the chemical was more harmful to marine life than the oil.

In the vicinity of the thickest portion of the slick, the operator positioned a "sea curtain"—a plastic boom—and attempted to vacuum up as much of the oil as possible. To augment this effort, a huge barge was outfitted to skim oil from the surface. The barge's equipment included three water knockouts, designed to separate as much water as possible from oil. The barge mounted two tanks, providing a tank capacity of approximately 40,000 barrels, and required three tugs for its skimming chore, including two to pull it forward and one behind to provide the proper tensioning effect.

The most effective clean-up method consisted of spraying straw to absorb oil. Four straw-spreaders, mounted on workboats, sprayed straw along the coastline at a distance of several hundred yards from shore, attempting to absorb the oil before it reached the sand, making it possible for other vessels to "harvest" the oil-soaked straw. The straw-spreaders were the type normally used for spreading straw on open cuts along new highways to prevent soil erosion. Straw also was spread on beaches to absorb oil.

After the well was controlled, beach clean-up was accelerated. Supplementary manpower, obtained from the State Department of Conservation, was augmented with prison and contract labor, bringing the total to a peak of nearly 1,000 men. The clean-up of beaches was accomplished primarily by hand raking up of oil-soaked straw.

Disposal of the straw posed a problem. Santa Barbara city officials denied use of the city dump. The Santa Barbara County dump was inaccessible because of a washed-out road. It was necessary to haul debris to the Ventura and Oxnard dumps, and some loads were hauled as far as Fillmore, a distance of over 50 miles.

There was bitter controversy over how much—and how permanently—the channel might have been damaged by the oil spill.

Even as newspapers carried accounts of stricken seals and whales—and some voices rose to refute the charges—a crowd variously estimated at from 400 to 700 gathered on Sunday afternoon, April 6, for an anti-oil rally at the foot of Stearns Wharf.

The crowd listened to speeches, then part of the group moved onto the wharf to stage a sit-down that turned back four service and supply trucks seeking access to installations on the wharf. Chanting "Get oil out" and "Spirit of '76," members of the crowd wrote their names and smeared grease on cranes operated by Standard Oil Company of California at the end of the wharf.

Two nights later, a crowd of more than 300 jammed a Santa Barbara city council meeting, demanding that the wharf be closed to oil support activities. After a stormy meeting, the mayor and councilmen walked out and members of the audience went to the rostrum to assume the vacated chairs. City Attorney Stanley Tomlinson said, "It was pure chaos, pure anarchy, with everyone yelling at the same time. They demanded immediate action on something that can only be done with deliberation." The crowd left after police moved into the chamber. There were no arrests.

Some voices rose to the defense.

Frank J. Hortig, executive officer of the State Lands Commission, pointed out at the commission's February meeting that more than 1.5 million feet of hole had been

A massive 50-ton blowout preventer aboard the WODECO IV prior to installation on the ocean floor 1,000 feet below. The rams could close on anything in the hole, or on themselves, to prevent a well from getting away. (William Rintoul)

cut since 1955 on the state's offshore leases "without a single drop of pollutant getting into the Pacific Ocean." Hortig said some 925 wells had been drilled from offshore platforms and islands.

In a letter to the editor of the *Santa Maria Times*, William C. Wilbur Jr., Santa Maria, wrote: "I wonder, Mr. Editor, how willing you would be to cease using all petroleum products for the duration of your proposed drilling shutdown. Would you do without gasoline and oil for your car and walk to wherever you had to go? Would you eat uncooked meals in an unheated home to avoid use of natural gas? To oppose the cause while enjoying the effect is to be guilty of hypocrisy."

In the columns of the *Los Angeles Times*, another letterwriter, Robert K. Kassenbrock, Whittier, said: "That the recent offshore oil well blowout was an unfortunate happening is beyond question. The *Times* in its reporting, editoralizing and cartooning of the event was seriously one-sided. More concern was evidenced for marine and bird life than for the lives of the brave men who were battling to bring the well under control.

"These are the same men who daily risk their lives in the quest for offshore oil so that California, the largest gasoline market in the world, will not be dependent on foreign imports. The offshore oil industry has paced the United States in the technological lead in underseas exploration, spawned many new companies employing thousands and paying billions in revenues to local, state and federal governments.

"The recent disaster at International Airport where lives were lost brought no cries for the shutdown of the airport or the grounding of aircraft by the FAA. Why then the exhortation to shut down offshore oil operations, putting thousands out of work?

"Let's keep happenings in their proper perspective and realize that the community and industry must coexist for the common good."

Another *Times* reader, John A. Clemente, Santa Monica, noting common reference to an 800-square-mile oil slick— the figure most often used by the *Times* and other publications, wrote: "An 800 square mile oil slick with an average thickness of only one-tenth of an inch would contain 33,100,800 barrels of oil (Yes! Thirty-three million.) Is there a possibility that the public will be provided with some realistic figures for the true volume of oil which was dumped into the Santa Barbara Channel?"

In its wake, the blowout left a legacy of cancelled lease sales, lawsuits and proposed legislation to halt offshore drilling.

The state of California postponed "indefinitely" sales that had been planned for the Santa Barbara Channel and for San Pablo Bay some 16 miles northeast of San Francisco. Before the latter sale was cancelled, a self-propelled barge christened the *Bakersfield* and its jug boat had shot the waters of San Pablo Bay for a group of companies including Mobil Oil Corporation, Signal Oil & Gas Company, Tenneco Oil Company, Texaco Inc. and Union Oil Company of California. It was rumored that the seismic survey had turned up more than one structure.

The federal government cancelled plans for a sale of leases in San Pedro Bay, which tentatively had been scheduled for 1970, postponed a drainage sale that had been scheduled off Louisiana, and delayed other sales that were pending off Texas, western Louisiana and in the Gulf of Alaska.

Four major lawsuits were initiated. On the day the blowout was controlled, five individuals suing on behalf of all damaged parties filed a $1.3 billion damage suit in Santa Barbara Superior Court against Union and its partners, Mobil, Gulf and Texaco.

Also in Santa Barbara, Charles A. O'Brien, California chief deputy state attorney general, filed a $560 million suit against Union and its partners on behalf of all public agencies and entities affected by the oil spill.

The County and City of Santa Barbara filed suit in U.S. District Court in Los Angeles seeking an injunction to halt drilling on federal leases "permanently and forever."

Six independent oil companies headed by Pauley Petroleum Inc. filed suit in the U.S. Court of Claims in Washington, D.C., for $230 million compensation for what they described as Interior Secretary Hickel's interference with their drilling operations in the channel. The petition said Hickel's promulgation of a new regulation imposing on the lessee absolute liability, regardless of fault, for any escape of oil "rendered it economically and practically impossible for plaintiffs to continue further drilling or production operations."

Legislation proposed in Washington included a measure offered by Senator Alan Cranston (Democrat-California) that would terminate all drilling on federal leases in the channel and suspend operations on all other Outer Continental Shelf leases off California and another offered by Representative Teague that would have swapped Santa

Barbara Channel leases for Naval Petroleum Reserve No. 1 at Elk Hills, 18 miles southwest of Bakersfield in the San Joaquin Valley.

Meanwhile, Humble Oil & Refining Company, when permitted to resume work in the spring of 1969, continued with exploration of the deep-water tracts it had acquired in the February, 1968, sale in what the company's geologists referred to as the Five-Mile Trend about 20 miles west of Santa Barbara.

Before the end of 1970, the company had confirmed discovery of three fields. The firm did not publicly set a figure on how much oil it had found, but sources outside the company estimated that the discoveries represented one billion or more barrels of oil.

In the 10 years that followed, Humble and its successor, Exxon Company U.S.A., in an effort to produce oil from the Santa Barbara Channel tracts would run a bureaucratic gauntlet that included three major environmental impact studies, 21 major public hearings, 10 major governmental approvals and many more minor ones, 51 studies by consultants costing more than $2 million, 12 lawsuits and one Santa Barbara County referendum, which the company won. The company would not succeed in producing a single barrel of oil.

Storm clouds over Platform A, foreground, site of the Santa Barbara Channel blowout, and WODECO II, on station for the drilling of a relief hole, portended troubled times in the wake of the Channel oil spill. (William Rintoul)

Index